QUANTITATIVE CHARACTERIZATION OF LIGAND BINDING

DONALD J. WINZOR
Department of Biochemistry
University of Queensland

WILLIAM H. SAWYER
Department of Biochemistry
University of Melbourne

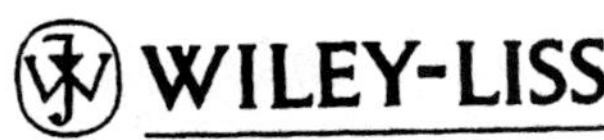
A JOHN WILEY & SONS, INC., PUBLICATION
New York • Chichester • Brisbane • Toronto • Singapore

Address All Inquiries to the Publisher
Wiley-Liss, Inc., 605 Third Avenue, New York, NY 10158-0012

A NOTE TO THE READER:
This book has been electronically reproduced from digital information stored at John Wiley & Sons, Inc. We are pleased that the use of this new technology will enable us to keep works of enduring scholarly value in print as long as there is a reasonable demand for them. The content of this book is identical to previous printings.

Library of Congress Cataloging-in-Publication Data

Winzor, Donald J.
Quantitative characterization of ligand binding / by Donald J. Winzor and William H. Sawyer.
p. cm.
Includes bibliographical references and index.
ISBN 0-471-05957-9 (hardcover : alk. paper). — ISBN 0-471-05958-7 (paperback : alk. paper)
1. Ligand binding (Biochemistry) I. Sawyer, William H. II. Title.
QP517.L54W54 1995
574.19'283—dc20 95-31858
CIP

The text of this book is printed on acid-free paper.

QUANTITATIVE CHARACTERIZATION OF LIGAND BINDING

CONTENTS

PREFACE

The specific binding of one molecule to another is a phenomenon ubiquitous to all fields of biology. Specific recognition and binding processes control the extent and frequently the rate of biological reactions. The biochemist has the task of evaluating a myriad of ligands, substrates, cofactors, inhibitors, and effectors that bind specifically to protein/enzyme acceptor molecules. The biochemist also has the task of assessing their importance *in vivo*. Antigen–antibody interactions are the lifeblood of immunology. Quantitative characterization of the maze of interactions involving proteins and nucleic acids remains an essential though largely unexplored area of research for the molecular biologist. Signal transduction through biomembranes culminates in a cascade of events involving recognition and binding of cellular components. Interactions of drugs with protein carriers and also with their ultimate receptor targets are of pivotal concern to the pharmacologist, who must also deal with the additional complexity introduced by the binding of a ligand to an insoluble acceptor matrix such as a biomembrane. In this respect, the interactions of proteins, hormones, or drugs with matrix proteins and membrane receptors hold the key to our understanding of cellular events such as muscle contraction, hormone responses, and lymphocyte activation. Binding processes are the essence of biological control, and their quantitative characterization is therefore essential for an understanding of intracellular and extracellular events in molecular terms.

By spreading its impact across such a broad spectrum of research areas, ligand binding is a phenomenon that has shown scant disregard for the current practice of dividing biology into clearly defined disciplines. The equilibrium nature of ligand-binding interactions necessitates their quantitative study by specialized techniques, the development of which has been largely the province of physical biochemists. Accordingly, most reviews of ligand binding have concentrated on mathematical aspects of the procedures, an emphasis that can prove rather daunting to, say, a budding neurochemist or membrane biologist. The aim of the present book is to transgress the traditional interdisciplinary barriers by presenting the methodological and interpretative aspects of ligand-binding studies in a manner designed to serve as a useful introduction of the subject to biology graduates/undergraduates, and, indeed, to any biologist contemplating an initial foray into the quantitative characterization of a ligand-binding phenomenon.

DONALD J. WINZOR
WILLIAM H. SAWYER

Brisbane and Melbourne

1

INTRODUCTION: THE ANALYSIS OF SIMPLE BINDING RESPONSES

Biological systems abound with interactions involving the binding of a small solute to a macromolecule, be it protein, polysaccharide, or polynucleotide. Such binding can provide the means of solute transport within the physiological system, as exemplified by the oxygen–hemoglobin, iron–transferrin, and drug–albumin interactions. Alternatively, it can provide a controlling influence over physiological function. For example, the reversible binding of Ca^{2+} to proteins plays a crucial role in the control of muscle action and blood coagulation. Similarly, the regulation of metabolic pathways entails an interplay of equilibria involving enzymes and metabolites. To this list of biologically significant binding phenomena may also be added the immunological protection afforded by antibody–antigen interactions; the controls of gene expression and repression, of hormone action, and of neuronal responses; and the pharmacological benefit of interactions between drugs and their specific receptors.

A biologically important feature of these binding phenomena is the noncovalence of the interactions, the equilibrium nature of which thus allows the extent of complex formation to vary with prevailing reactant concentrations and hence with fluctuations in the metabolic state of the cell. Because this composition dependence of the extent of reaction is governed by the equilibrium constant for the interaction, the usual goal of binding studies is the determination of this equilibrium constant to allow the distribution of the two reactants between free and complexed states to be predicted for any specified combination of the two reactant concentrations. Binding studies may also be used to comment on the nature of the bond responsible for the noncovalent interaction. These more extensive binding studies require delineation of the temperature dependence of the equilibrium constant (K) and hence calculation of the standard enthalpy ($\Delta H°$) and the corresponding change in entropy ($\Delta S°$) from the fundamental thermodynamic expressions

$$d\,(\ln K)/dT = \Delta H^\circ/RT^2 \tag{1.1}$$

$$\Delta G^\circ = -RT \ln K = \Delta H^\circ - T\Delta S^\circ \tag{1.2}$$

A negative ΔH° implies the existence of a favorable enthalpic contribution to the standard free energy (ΔG°) and often implicates hydrogen bonding as a substantial contributor to the strength (ΔG°) of the interaction. On the other hand, predominance of hydrophobic and/or electrostatic factors leads to the situation in which the size of a positive entropic change (ΔS°) dictates the magnitude of ΔG°. Under these circumstances the binding constant increases with increasing temperature (eq. 1.2). Distinction between hydrophobic and electrostatic effects as the major influence in an entropically driven equilibrium reaction is made on the basis that K for the latter type of interaction exhibits an inverse dependence on ionic strength.

The equilibrium nature of these binding phenomena necessitates their quantitative study by specialized techniques, since the composition of the equilibrium mixture is likely to be perturbed by any attempt to remove the complex or either reactant for determination of its concentration in the mixture. Chapters 2–5 are devoted to the description of these specialized procedures, whereas the emphasis in Chapters 6 and 7 is on the analysis and interpretation of results so obtained—a topic that is brought into this introductory chapter by consideration of the simplest ligand-binding systems. Chapter 8 addresses the study of DNA-ligand interactions in the light of all the preceding considerations. As a prelude to those endeavors, it is necessary to define the problem of characterizing ligand binding in general terms.

1.1 DEFINITION OF TERMS IN LIGAND BINDING

In studies of the noncovalent interaction between two species, the smaller solute is referred to as the *ligand*, which we designate as S and consider to possess a single site for interaction with the larger species. The latter is termed the *acceptor* and is accordingly designated as A in systems where the larger solute is soluble. However, in instances where the interaction is between ligand in solution and a particulate acceptor, the term *receptor* is often used to describe A. Whereas the ligand is generally considered to be univalent in its interaction with A, the acceptor may possess several sites for interaction with S and is therefore said to be *multivalent*. Since the designation of the two interacting solutes as *small* and *large* clearly involves operational definitions, there is no impediment to the same solute being regarded as the acceptor in one system but as the ligand in another. For example, in the interaction of lysozyme with *N*-acetylglucosamine the enzyme would be regarded as the acceptor; but in a study of the interaction between lysozyme and a cell-wall preparation, the enzyme would be the ligand. A similar switch in nomenclature would apply to the interactions of glyceraldehyde-3-phosphate dehydrogenase with NADH and with either muscle myofibrils or erythrocyte ghosts; but in the latter cases multivalence of the enzymatic ligand is a complicating (but not insuperable) factor in the analysis (see Chapter 7).

Even for a system involving the reversible binding of ligand, S, to p sites on acceptor, A, the concentrations of p complexes, *viz.*, AS, AS_2, . . . , AS_p, as well as those of free acceptor and free ligand, are required to define completely the composition of any given acceptor–ligand mixture. From an experimental viewpoint it would be unduly optimistic to think that any analytical procedure might provide these $(p + 2)$ concentrations directly. A more realistic outlook is that an experimental procedure is available for determining either C_S, the molar concentration of free ligand, or $(\overline{C}_S - C_S)$, the concentration of ligand present as acceptor–ligand complexes, in a mixture with defined total concentrations $\overline{C}_A$ and $\overline{C}_S$ of acceptor and ligand, respectively. Any further analysis of the system must therefore be contemplated with this restriction in mind.

1.2. THE BINDING FUNCTION

In studies of mixtures of acceptor A and ligand S, it is possible to define operationally a binding function, r (Klotz, 1946), also termed ν (Scatchard, 1949), as the molar ratio of the amount of ligand bound to the total amount of acceptor. On taking into account the fact that both of these amounts are contained in the same volume, it follows that

$$r = (\overline{C}_S - C_S)/\overline{C}_A \tag{1.3}$$

where C_S denotes the equilibrium concentration of free ligand in a mixture with total molar concentrations $\overline{C}_S$ and $\overline{C}_A$ of ligand and acceptor, respectively. Since this function corresponds to the average number of molecules associated with each acceptor molecule, it follows that its value is within the range $0 \leq r \leq p$ for an acceptor with p sites for interaction with ligand. In some investigations results are reported in terms of the fractional saturation of acceptor sites, f_a, which, being defined as r/p, has limiting values of zero and unity. Use of f_a rather than r does, however, rely upon knowledge of p, or upon the expression of results in terms of some parameter related directly to p.

1.3. THE PROBLEM: DETERMINATION OF BOUND OR FREE LIGAND

Provided that the total concentrations of acceptor and ligand $(\overline{C}_A, \overline{C}_S)$ are known or can be determined, the evaluation of r clearly requires the measurement of either C_S, the concentration of free ligand, or $(\overline{C}_S - C_S)$, the concentration of ligand bound. The latter quantity, or a property related thereto, is available for some systems by application of spectral techniques, which are the subject of Chapter 3, and for others by use of biosensor technology (Chapter 5). More frequently, however, it is C_S that is determined—by techniques described in Chapters 2 and 4.

In the determination of the binding function by measurement of C_S or $(\overline{C}_S - C_S)$ it is imperative that the method used does not perturb the equilibrium. Extreme care should thus be exercised in the use of radioimmunoassay procedures based on measurement of C_S after precipitation of all antigen–antibody complexes by the addition of ammonium sulfate. Only in the event that the rates of complex formation and dissociation are very slow does the concentration of free ligand in the supernatant correspond to that in the equilibrium mixture prior to ammonium sulfate addition. On the other hand, rapid association/dissociation could well lead to the situation wherein the concentration of ligand measured refers to the new equilibrium position effected by the dramatic change in ionic strength.

Likewise, the validity of the commonly used nitrocellulose filter assays for the study of protein–nucleic acid interactions, and of many solid-phase radioimmunoassay and ELISA systems, should be subjected to close scrutiny. These all rely upon measurement of the concentration of complexed ligand associated with the solid phase after removal of uncomplexed ligand by rinsing of the filter/plate with buffer. For any reliance to be placed on results obtained by these procedures it is clearly necessary to demonstrate that r is independent of the buffer volume used to remove free ligand. The fact that the use of cold buffer is usually recommended for the rinsing process could well indicate that the assumed negligibility of complex dissociation during the removal of free ligand is a questionable approximation. Although it is true that the rate of complex dissociation would be lowered by the decrease in temperature, such a course of action also introduces the possibility that the equilibrium position may be perturbed in accordance with the Gibbs-Helmholtz relationship (eq. 1.2).

From the above general discussion of the problem it should be evident that correct experimental protocol can give rise to an estimate of r for a mixture with defined values of C_S and $\overline{C}_A$. A series of such experiments with the same $\overline{C}_A$ but a range of $\overline{C}_S$ then allows the construction of a binding curve, *i.e.*, a plot of the dependence of r upon C_S. For many systems repetition of the experiments with a different total acceptor concentration yields the same binding curve; but for acceptors and/or ligands undergoing self-association, a different binding curve pertains to each acceptor concentration (Chapter 6). Since variation of $\overline{C}_A$ for the latter types of system would lead to the construction of a binding curve that is a meaningless composite of those for the acceptor concentrations used, the safest course of action is clearly to determine any binding curve at a fixed total concentration of acceptor.

1.4. RECTANGULAR HYPERBOLIC BINDING RESPONSES

Irrespective of the procedure that is used to obtain a binding curve (r *vs.* C_S), the next step is its interpretation in terms of the binding equation for the appropriate model of the acceptor–ligand interaction. For that purpose we need to introduce the various ways of defining the equilibrium constant for an acceptor–ligand interaction that conforms with the simplest binding behavior—the rectangular hyperbolic response.

Consider initially the situation where ligand binds to p sites on a single acceptor state, for which the successive equilibria are described in terms of molar association constants, K_i, by

$$\begin{aligned} A + S &\leftrightarrows AS; & K_1 &= C_{AS}/(C_A C_S) \\ AS + S &\leftrightarrows AS_2; & K_2 &= C_{AS_2}/(C_{AS} C_S) \\ \cdots &\;; & &\cdots \\ \cdots &\;; & &\cdots \\ AS_{p-1} + S &\leftrightarrows AS_p; & K_p &= C_{AS_p}/(C_{AS_{p-1}} C_S) \end{aligned} \tag{1.4}$$

By successive substitution in eq. 1.4 it follows that the concentration of each acceptor–ligand complex may be expressed in terms of the free concentrations of acceptor and ligand as

$$C_{AS_i} = \left(\prod_1^i K_i\right) C_A C_S^i \tag{1.5}$$

where the term in parentheses denotes the product of the association constants for the successive equilibrium reactions leading to the formation of AS_i from A and S. Because the total concentrations of acceptor and ligand may therefore be written in the form

$$\overline{C}_A = C_A + \sum_1^p C_{AS_i} = C_A\left[1 + \sum_1^p i\left(\prod_1^i K_i\right) C_S^i\right] \tag{1.6a}$$

$$\overline{C}_S = C_S + \sum_1^p iC_{ASi} = C_S + C_A \sum_1^p i\left(\prod_1^i K_i\right) C_S^i \tag{1.6b}$$

the binding equation (eq. 1.3) becomes

$$r = \sum_1^p i\left(\prod_1^i K_i\right) C_S^i \Bigg/ \left[1 + \sum_1^p \left(\prod_1^i K_i\right) C_S^i\right] \tag{1.7}$$

which is the expression proposed originally by Adair (1925).

Although eq. 1.7 provides, in principle, an unequivocal description of binding data, its application requires the elucidation of $p + 1$ parameters, the stoichiometery (p) as well as the p stoichiometric association constants (K_i), by polynomial curve fitting of the binding curve (r *vs.* C_S). A further disadvantage of this thermodynamically rigorous approach is that description of the binding function as a ratio of polynomials gives little insight into the nature of the interaction being studied. In the belief that the aim of most experimental binding studies is to shed light on the numbers and affinities of acceptor sites for ligand, we adopt the less rigorous approach whereby the thermodynamic description is based upon a specific model of the interaction. Such action, which brings studies of ligand binding more into line with their enzyme kinetic counterparts, means that the outcome of a binding study is an illustration of the degree of conformity of the experimental data with thermodynamic description in terms of the selected model of the interaction. However, that criticism pervades any attempt to ascribe molecular significance to thermodynamic parameters, which, being only dependent upon a difference between initial and final states, are necessarily independent of the pathway (mechanism) by which thermodynamic equilibrium was attained from the initial state.

Considerable simplification of eq. 1.7 is achieved (Klotz, 1946) if all stoichiometric constants (K_i) can be related to a single intrinsic, or site-binding, constant, k_{AS}. Thus, provided that the interactions of ligand with acceptor sites are all equivalent and occupancy independent, the intrinsic association constant, k_{AS}, is related to the stoichiometric constant, K_i, by

$$K_i = (p - i + 1)k_{AS}/i \tag{1.8}$$

Subject to the validity of applying the above restrictions of equivalence and independence of sites, eq. 1.7 simplifies to

$$r = \frac{pk_{AS}C_S(1 + k_{AS}C_S)^{p-1}}{(1 + k_{AS}C_S)^p} = \frac{pk_{AS}C_S}{1 + k_{AS}C_S} \tag{1.9a}$$

or, if the equilibrium is described in terms of an intrinsic dissociation constant, $k_d = 1/k_{AS}$,

$$r = pC_S/(k_d + C_S) \tag{1.9b}$$

Such rectangular hyperbolic dependence of r upon C_S is frequently examined by means of one of the following linear transforms of eq. 1.9:

$$r/C_S = pk_{AS} - rk_{AS}; \quad \text{Scatchard plot} \tag{1.10a}$$

$$1/r = 1/(pk_{AS}C_S) + (1/p); \quad \text{Double-reciprocal plot} \tag{1.10b}$$

$$C_S/r = 1/(pk_{AS}) + (1/p)C_S; \quad \text{Hames plot} \tag{1.10c}$$

Illustrative plots of the rectangular hyperbolic dependence and its three linear transforms are presented in Figure 1.1 for the binding of NADH to lactate dehydrogenase (Ward and Winzor, 1983).

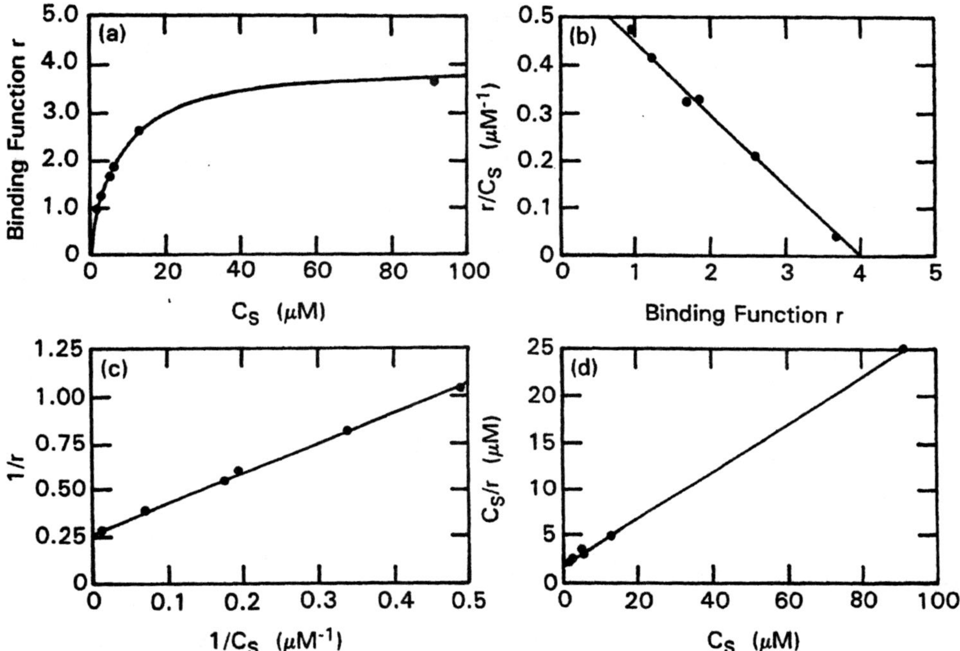

Fig. 1.1. Various methods of presenting data for the interaction of NADH with rabbit muscle lactate dehydrogenase (Ward and Winzor, 1983): **(a)** untransformed plot; **(b)** Scatchard plot (eq. 1.10a); **(c)** double-reciprocal plot (eq. 1.10b); **(d)** Hames plot (eq. 1.10c).

Although many interactions conform with this simplest model of ligand binding, there are also many acceptor–ligand interactions for which its requirement of equivalent and independent binding sites on acceptor does not apply. For example, sites may deviate from this restrictive combination of conditions by virtue of their nonequivalence. Alternatively, the affinity of a given acceptor site for ligand may depend upon the occupancy of other acceptor sites. At this stage we shall restrict attention to the former case and leave the consequences of cooperative binding until Chapter 6.

In the event that a ligand binds to two different classes of independent sites on an acceptor, the binding equation becomes the sum of the rectangular hyperbolic expressions describing the interaction of ligand with each class of sites. Thus,

$$r = \{p(k_{AS})_1 C_S/[1 + (k_{AS})_1 C_S]\} + \{q(k_{AS})_2 C_S/[1 + (k_{AS})_2 C_S]\} \quad (1.11)$$

where p and q denote the respective numbers of acceptor sites for which the interaction with ligand is governed by intrinsic binding constants $(k_{AS})_1$ and $(k_{AS})_2$. The presence of further classes of binding sites on the acceptor is readily accommodated by the inclusion of additional rectangular hyperbolic expressions

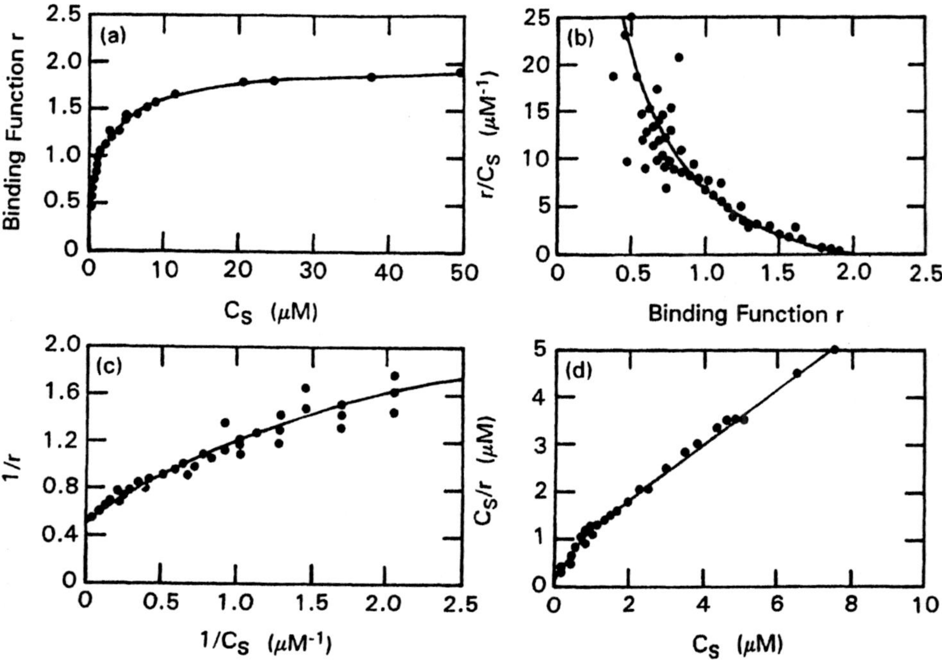

Fig. 1.2. Results for the interaction of dicoumarol with human serum albumin (Perrin *et al.*, 1975), showing the effect of acceptor-site heterogeneity on the forms of binding curves: **(a)** untransformed data; **(b)** Scatchard plot; **(c)** double-reciprocal plot; **(d)** Hames plot. The corresponding plots for a system with a single class of binding sites are shown in Figure 1.1.

on the right-hand side of eq. 1.11. Nonequivalence of acceptor sites for ligand may be recognized by plotting results in accordance with any of the linear transforms of the rectangular hyperbolic binding expression (eqs. 1.10a–c). Scatchard, double-reciprocal, and Hames plots assume curvilinear form, as is evident from such analyses (Fig. 1.2) of the binding of dicoumarol to two different sites on human serum albumin with intrinsic binding constants of 2.9 $\times$ 10^6 M^{-1} and 1.9 $\times$ 10^5 M^{-1} (Perrin *et al.*, 1975).

1.5. QUANTITATIVE EVALUATION OF BINDING PARAMETERS

From Figure 1.1 and eqs. 10a–c it is evident that any linear transform of the rectangular hyperbolic expression provides a simple means of characterizing the binding of ligand to a single class of acceptor sites. Thus, the slope and ordinate intercepts of a linear Scatchard plot are $-k_{AS}$ and pk_{AS}, respectively, whereas the corresponding parameters from a double-reciprocal plot yield $1/pk_{AS}$ and $1/p$. The Hames plot has a slope of $1/p$ and an ordinate intercept of $1/pk_{AS}$. However, an undesirable consequence of using eqs. 1.10a–c for such characterization of ligand binding is the distortion of experimental uncertainty effected by the linear transformation (Klotz, 1983). This aspect of parameter evaluation from indirect plots becomes clear in Figure 1.2, where the large differences in data placement and extents of experimental scatter are manifestations of the bias introduced by analysis in terms of a linearly transformed version of the basic binding equation.

Nonlinear regression analysis of the untransformed (r, C_S) data in accordance with the appropriate binding expression (eq. 1.9 or 1.11) is the preferred method for evaluating binding parameters, there being many such computer programs currently available. However, the application of such procedures relies upon the provision of initial estimates of parameters. Because the human eye can detect more readily any departure from linearity than a subtle change in curvilinear form, plots such as those presented in Figures 1.1 and 1.2 identify the appropriate binding expression (eq. 1.9 or 1.11) for nonlinear regression analysis and also provide the required initial estimates of binding parameters. Consequently, despite their inevitable distortion of the consequences of experimental uncertainty, the linear transforms of the rectangular hyperbolic binding equation continue to serve useful illustrative and diagnostic roles, even though the ultimate characterization of the binding phenomenon ensues from less biased analysis of the untransformed (r, C_S) data. Such resort to nonlinear regression analysis overcomes the problem that the abscissa intercept of a Scatchard plot may give an underestimate of p—a point raised by Klotz (1982), who favors a semilogarithmic presentation of binding data (r *vs.* log C_S) to provide a better indication of the proportion of the binding curve covered by the set of experimental results.

1.6. RELATIONSHIPS BETWEEN LIGAND BINDING AND ENZYME KINETICS

Although not usually regarded in such terms, enzyme kinetic studies may often be considered as a form of ligand-binding estimation in the sense that the initial velocity (v) provides a measure of ($\bar{C}_S \approx C_S$), and the maximal velocity (v_m) a corresponding measure of $p\bar{C}_A$. Since the ratio v/v_m is equivalent to r/p, enzyme kinetic studies share with their spectral counterparts the ability to provide the fractional saturation parameter, f_a, as a function of substrate (ligand) concentration, C_S (Frieden, 1964). However, construction of the enzyme kinetic plot in terms of total substrate concentration (v *vs.* C_S or one of its linear transforms) does not violate this proposition, because the approximations $\bar{C}_S >> \bar{C}_A$ (where A denotes enzyme) and hence $\bar{C}_S \approx C_S$ are inherent in treatments of results from initial velocity studies for small substrates.

The above inference that enzyme kinetic studies define the form of the binding curve for substrate does, of course, identify K_m for an enzyme exhibiting Michaelis-Menten kinetics, namely,

$$A + S \underset{k_2}{\overset{k_1}{\leftrightarrows}} AS \overset{k_3}{\rightarrow} A + P \tag{1.12}$$

as a thermodynamic parameter ($k_d = k_2/k_1$). This concept of the Michaelis constant as a dissociation constant is in keeping with the original enzyme kinetic analysis based on rapid equilibrium attainment (Michaelis and Menten, 1913), but is at variance with the steady-state concept of Briggs and Haldane (1925), who favor consideration of K_m as the ratio of kinetic constants $(k_2 + k_3)/k_1$. Clearly the two treatments are equivalent if $k_3 << k_2$. Even if the steady-state situation pertains, there are many circumstances under which enzyme kinetic results are equivalent to the rapid equilibrium case, except that the Michaelis constant so determined is the Briggs-Haldane ratio of rate constants rather than the dissociation constant (Hearnon *et al.*, 1959; Frieden, 1964; Dalziel, 1968; Kuchel *et al.*, 1974).

1.7. A STRATEGY FOR BINDING STUDIES

The fact that this introductory chapter has concentrated on rectangular hyperbolic binding responses is not intended to convey the impression that the characterization of ligand binding merely entails the evaluation of two parameters—the intrinsic binding constant (k_{AS}) describing the interaction of a ligand with a certain number (p) of sites on the acceptor. More complicated behavior results when the binding of the ligand to the acceptor is cooperative, when the acceptor or ligand exists as an equilibrium mixture of conformationally distinct or oligomeric forms, when the ligand is multivalent, or when the ligand binds to a linear polymer such as DNA. Indeed, the number of possible models of an acceptor–ligand interaction can become somewhat daunting to the uninitiated. The classic approach to the analysis of experimental binding data is to formulate a model that is based on the data but that also takes into account any complementary information relevant to the system. For example, spectroscopic analysis may indicate the occurrence of ligand-induced conformational changes, or molecular weight measurements may signify self-association of the acceptor. A mathematical description of the model is then used to curve-fit the experimental results and to simulate data in order to appreciate more fully the relationships between reactant concentrations and binding parameters. Simulations thus act to prompt and guide the design of further experiments that will test more thoroughly the model being invoked to describe the currently available information. This strategy is depicted diagrammatically in Figure 1.3. In any attempted quantitative characterization of ligand binding the important point to remember is that agreement of experimental data with a theoretical model does not establish the mechanistic correctness of that model. As with kinetic studies, it merely signifies consistency between the two. Nevertheless, the observation

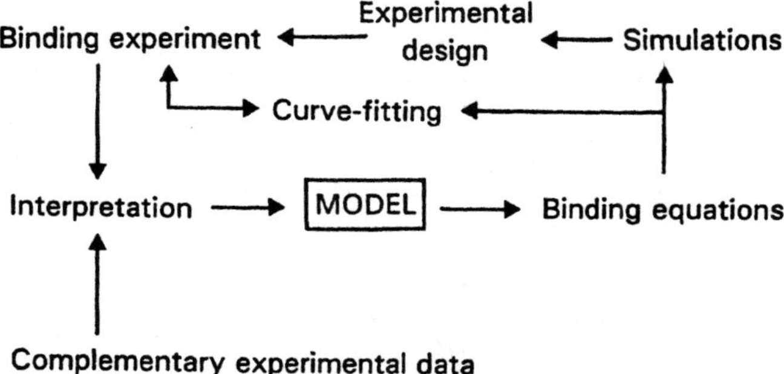

Fig. 1.3. Schematic representation of the strategy for the quantitative characterization of a binding response.

of such agreement between theory and experiment does fulfill the primary goal of binding studies, which is to obtain an adequate thermodynamic description of the interaction so that the evaluated parameters may be used to predict the composition dependence of binding responses under the conditions to which the characterization refers.

2

EXPERIMENTAL METHODS BASED ON PHASE SEPARATION

The more general procedures for evaluating the concentration of free ligand in acceptor–ligand mixtures entail the creation of a solution phase containing only the ligand, present at a concentration that may be related to its equilibrium counterpart in the equilibrium mixture. In this chapter the emphasis is placed on methods that yield C_S by such means.

2.1. EQUILIBRIUM DIALYSIS AND STEADY-STATE DIALYSIS

2.1.1. Equilibrium Dialysis

In this technique a membrane that is permeable to ligand but not to acceptor or acceptor–ligand complexes is used to divide the dialysis cell into two compartments (Fig. 2.1). After introducing a solution of acceptor or of acceptor–ligand mixture into one compartment (α) and ligand or buffer into the other (β), the system is allowed to equilibrate at constant temperature until thermodynamic equilibrium has been attained. Depending on the size and design of the dialysis cell, the period taken to achieve dialysis equilibrium ranges from several hours to several days. At equilibrium the thermodynamic activity of ligand in the β-compartment (*i.e.*, the pure-ligand phase) equals that of free ligand in the mixture. This means that in the absence of thermodynamic nonideality or charge effects, the concentration of ligand in the β-compartment defines its equilibrium concentration in the mixture ($C_S^\beta = C_S^\alpha$). Because the β-compartment contains only free ligand, there are numerous procedures that may be used for its estimation, enzymatic, spectrophotometric, or radiochemical tracer procedures being most commonly used.

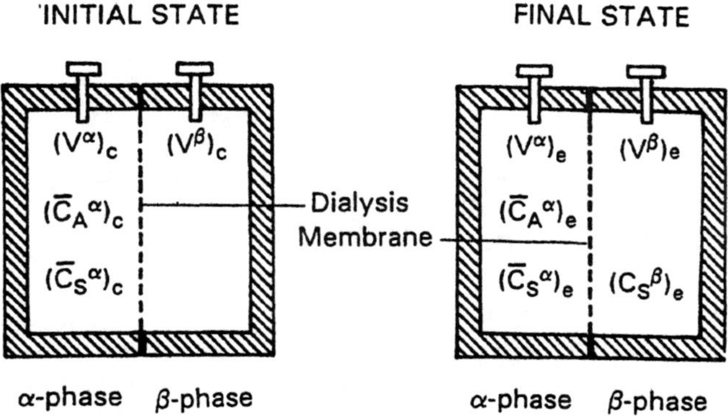

Fig. 2.1. Essential features of the equilibrium dialysis method for characterizing acceptor–ligand interactions, together with a summary of present nomenclature for use with eqs. 2.1–2.5.

The chief problem in equilibrium dialysis is the need to determine $\bar{C}_S^\alpha$, the total ligand concentration in the mixture, which clearly differs from its initial value due to transfer of ligand between compartments to achieve dialysis equilibrium. A second complication is that the total acceptor concentration may also differ from its initial value as the result of net osmotic flow of liquid between the two phases. Of the two approaches that are presented to overcome these problems, the first applies to situations where $\bar{C}_S^\alpha$ and $\bar{C}_A^\alpha$ are directly measurable. The second is more general in that it applies to instances where neither of these quantities is available to the experimenter.

1. In the event that an analytical procedure such as scintillation counting or γ-counting is being used for the determination of ligand concentrations, the magnitudes of $\bar{C}_S^\alpha$ and C_S^β are obtained directly from analysis of the mixture and pure-ligand phases, respectively, at equilibrium. Having thus obtained a value for the numerator of r (eq. 1.3) as $(\bar{C}_S^\alpha - C_S^\beta)$, there only remains the problem of defining $\bar{C}_A^\alpha$, the total acceptor concentration in the equilibrium mixture. This quantity may be identified with the concentration of acceptor introduced into the α-compartment if there has been no net transfer of liquid from one compartment to the other; or it may often be obtained by direct analysis of the equilibrium mixture.

2. Should there be a significant change in the volumes of the two phases, or should it not be possible to measure the total ligand concentration in the mixture, a different procedure must be adopted. By way of illustration, let us consider that the experiment was commenced by placing a volume $(V^\alpha)_c$ of mixture with total concentrations $(\bar{C}_S^\alpha)_c$ and $(\bar{C}_A^\alpha)_c$ of ligand and acceptor, respectively, into the α-compartment of the dialysis cell and a volume $(V^\beta)_c$ of buffer in the other. At equilibrium, the total amount of ligand present must

clearly still be $(V^{\alpha})_c(\overline{C}_S^{\alpha})_c$; and accordingly we may write

$$(V^{\alpha})_c(\overline{C}_S^{\alpha})_c = [(V^{\alpha})_c + (V^{\beta})_c](C_S^{\beta})_e + (V^{\alpha})_e[(\overline{C}_S^{\alpha})_e - (C_S^{\beta})_e] \quad (2.1)$$

where the first term on the right-hand side accounts for free ligand distributed evenly throughout the total volume in the dialysis cell and the second term takes into account the equilibrium concentration of bound ligand, $[(\overline{C}_S^{\alpha})_e - (C_S^{\beta})_e]$, confined to a volume $(V^{\alpha})_e$, the volume in the α-compartment after attainment of dialysis equilibrium. Even in the absence of a value for this volume and/or concentration, eq. 2.1 may be used to evaluate their product, which is the amount of ligand bound. From the definition of r as the amount of ligand bound per mole of acceptor, it then follows that the binding function may be evaluated from the expression

$$r = \frac{(V^{\alpha})_c(\overline{C}_S^{\alpha})_c - [(V^{\alpha})_c + (V^{\beta})_c](C_S^{\beta})_e}{(V^{\alpha})_c(\overline{C}_A^{\alpha})_c} \quad (2.2)$$

since values of all parameters on the right-hand side of eq. 2.2 are available.

This indirect but more general procedure may well require quantitative correction for adsorption of ligand to the semipermeable membrane and/or cell assembly. To test for the necessity of so doing, several acceptor-free ligand solutions are subjected to the same dialysis regimen as that used for mixtures: the same values of $(V^{\alpha})_c$, $(V^{\beta})_c$, and a range of $(\overline{C}_S^{\alpha})_c$. If the application of eq. 2.1 to the results of these experiments yields a finite value for $(V^{\alpha})_e[(\overline{C}_S^{\alpha})_e - C_S^{\beta})_e]$, this amount of ligand is bound to the membrane or cell assembly and should be allowed for in the determination of r. In making such corrections it should be noted that the extent of adsorption is likely to vary with free ligand concentration and that the dependence of this correction factor upon $(C_S^{\beta})_e$ must be determined so that the appropriate value may be subtracted from the numerator of eq. 2.2 in the determination of r.

In applications of either procedure to systems involving the binding of an anionic or cationic ligand to an acceptor bearing net charge, it is necessary to consider the consequences of the Donnan redistribution of ions across a semipermeable membrane. In that regard the concentrations of a charged ligand in the two phases are related by the expression (Svensson, 1946)

$$[(C_S^{\alpha})_e/(C_S^{\beta})_e]^{1/Z_s} = 1 - [\overline{Z}_A(\overline{C}_A^{\alpha})_e/2I] \quad (2.3)$$

where $\overline{Z}_A$ and Z_S denote the net charges of acceptor (averaged over free and liganded states) and ligand, respectively, and $\mathbf{I}$ denotes the ionic strength of the buffer $[\mathbf{I} = (1/2)\ \Sigma\ (C_iZ_i^2)]$. In most studies performed at ionic strengths of 0.1 or higher, this Donnan correction is negligible; and only in studies of highly charged systems should there be any need to make allowance (via eq. 2.3) for the difference between $(C_S^{\alpha})_e$ and $(C_S^{\beta})_e$. Under those circumstances the statement of mass conservation for ligand (eq. 2.1) must be modified to

$$(V^{\alpha})_c(\overline{C}_S^{\alpha})_c = (V^{\alpha})_e(C_S^{\alpha})_e + (V^{\beta})_e(C_S^{\beta})_e + (V^{\alpha})_e[(\overline{C}_S^{\alpha})_e - (C_S^{\alpha})_e] \tag{2.4}$$

whereupon r is determined from the relationship

$$r = [(V^{\alpha})_c(\overline{C}_S^{\alpha})_e - (V^{\alpha})_e(C_S^{\alpha})_e - (V^{\beta})_e(C_S^{\beta})_e]/[(\overline{V}^{\alpha})_c(C_A^{\alpha})_c] \tag{2.5}$$

Provided that there is no significant extent of osmotic flow between the two phases, *i.e.*, that $(V^{\alpha})_c \approx (V^{\alpha})_e$, the value of r may thus still be determined. However, its evaluation clearly relies upon knowledge of Z_A, the charge on acceptor and, hence, via iterative determination of $\overline{Z}_A$, the average net charge on the acceptor in liganded and free states (eq. 2.6):

$$\overline{Z}_A = \left[\sum_0^p (C_{AS_i}^{\alpha}\, Z_{AS_i})\right] \Big/ \overline{C}_A^{\alpha} \tag{2.6}$$

2.1.2. Steady-State Dialysis

An obvious disadvantage of the previous procedure is the time required to achieve dialysis equilibrium across the semipermeable membrane, a factor that Colowick and Womack (1969) overcame by developing a steady-state dialysis procedure. The apparatus (Fig. 2.2a) comprises a dialysis cell with an upper compartment (α) containing the acceptor–ligand mixture, separated by a semipermeable membrane from a lower compartment (β), through which either buffer (Colowick and Womack, 1969) or recycled effluent (Ford *et al.*, 1984) is pumped at a constant flow rate. The liquid in both compartments is stirred to minimize concentration gradients within either phase. By analysis of the ligand concentration in the solution being continually removed from the lower compartment (C_S^{β}), it has been found that only a few minutes are required to establish a steady state between the rates at which ligand is entering and leaving this compartment (Fig. 2.2b). The resultant steady state in the effluent from the lower compartment is manifested either as a time-independent concentration of ligand (Colowick and Womack, 1969) or as a constant rate of concentration change in the method entailing the recycling of effluent (Ford *et al.*, 1984). Either steady-state parameter is then related to the concentration of free ligand in the upper compartment by calibrating the system with known concentrations of ligand (but no acceptor) in that compartment (Fig. 2.2c). To obtain binding data, acceptor–ligand mixtures with defined total concentrations of acceptor and ligand ($\overline{C}_A^{\alpha}$, $\overline{C}_S^{\alpha}$) are placed in the upper compartment, and the corresponding free ligand concentration (C_S^{α}) is determined on the basis of the resultant steady-state parameter and the linear calibration plot (Fig. 2.2c). The magnitude of the binding function then follows from eq. 2.1.

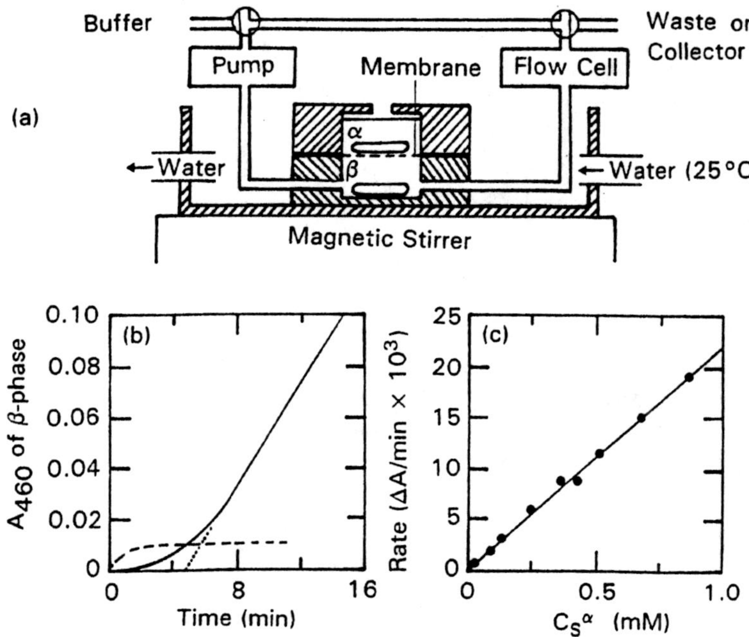

Fig. 2.2. Studies of ligand binding by steady-state dialysis. **(a)** Schematic representation of an apparatus designed to accommodate the original technique (Colowick and Womack, 1969) and a modification thereof involving recycling of the β-phase effluent (Ford *et al.*, 1984). **(b)** Time-course of the β-phase effluent absorbance in calibration experiments with 0.52 mM methyl orange for the original (dashed line) and recycling (solid line) versions of the technique, the linear limiting slope of the latter time-course being emphasized by its extension to the abscissa (dotted line). **(c)** Calibration plot for the evaluation of free methyl orange concentrations in albumin-dye mixtures. [Adapted from Ford *et al.* (1984).]

Despite the obvious advantage of the rapidity with which binding data may be obtained by this method, there have been some misgivings about use of the procedure because of the empiricism associated with its development. However, the theoretical basis of the technique has now been established (Ford *et al.*, 1984); and the approximation involved in consideration of the calibration plot to be linear has been identified as an assumed concentration independence of the diffusion coefficient for free ligand. Within the limits of accuracy with which C_S^α may be delineated, this assumption is an eminently reasonable approximation. Thus, although the steady-state dialysis procedure is empirical in the sense that the apparatus constants are defined operationally by means of a calibration plot, such empiricism should not be construed as uncertainty about the theoretical rigor of the method. It certainly does not constitute grounds for scepticism about the validity of results so obtained, except in instances where the presence of acceptor renders the acceptor–ligand mixtures much more viscous than the ligand solutions used for calibration purposes. From the Stokes-

Einstein relationship, the diffusion coefficient is inversely proportional to solution viscosity, and hence a calibration plot with slightly decreased slope would then need to be used to compensate for the decreased magnitude of the diffusion coefficient of ligand in mixtures. This modification of the calibration plot is best achieved through appropriate adjustment of the viscosity of the calibrating ligand solutions by inclusion of an inert polymer.

2.2. ULTRAFILTRATION AND DIAFILTRATION

Original versions of the ultrafiltration assembly (Fig. 2.3a) were designed for fixed-volume diafiltration (Blatt *et al.*, 1968), which entails the passage (under pressure) of ligand solution from a reservoir (chamber A) through a fixed volume of stirred acceptor solution (chamber B) until the ligand concentration in the effluent below the semipermeable membrane equals that (C_S^α) in the reservoir. The contents of chamber B are then assayed for total ligand concentration ($\bar{C}_S^\alpha$) to enable the binding function, r (eq. 1.3), to be calculated as $(\bar{C}_S^\alpha - C_S^\alpha)/\bar{C}_A^\alpha$, where $\bar{C}_A^\alpha$ is the total concentration of acceptor in the mixing vessel.

Much faster acquisition of binding data may be achieved (Paulus, 1969; Sophianopoulos *et al.*, 1978) by dispensing with the ligand solution in the reservoir and simply placing a mixture with known total concentrations of acceptor ($\bar{C}_A^\alpha$) and ligand ($\bar{C}_S^\alpha$) in chamber B. Application of pressure then allows the collection of ultrafiltrate at its equilibrium concentration (C_S^α) in the reaction mixture (Sophianopoulos *et al.*, 1978). Adoption of that experimental protocol has allowed values of r to be obtained within 10–20 minutes.

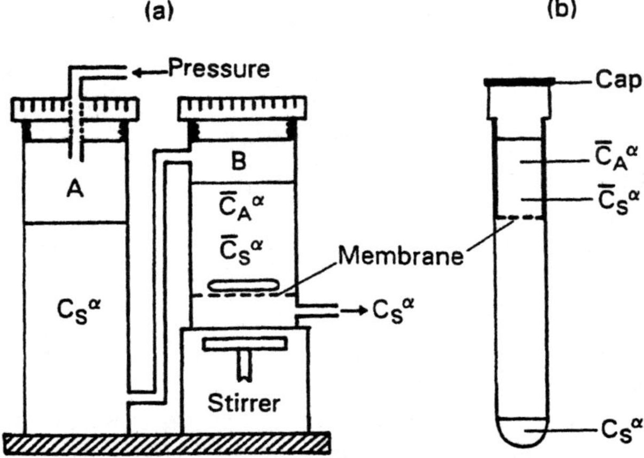

Fig. 2.3. Studies of ligand binding by ultrafiltration and diafiltration. **(a)** Schematic representation of the conventional apparatus in diafiltration mode. **(b)** Essential features of the centrifugal ultrafiltration technique.

Subsequent developments have been directed toward decreasing the volume of mixture (typically 5–6 ml in the above procedure) by employing centrifugation to generate the ultrafiltrate (Shah *et al.*, 1974; Hammond *et al.*, 1980). The reaction mixture (0.4 ml) is placed in a membrane-fitted sample cup that is inserted into the top of a centrifuge tube (Fig. 2.3b). Angular velocity (1,000–2,000 rpm) is then used to effect the field that forces ligand solution through the semipermeable membrane into the empty, lower half of the tube. The upper cup is then removed so that aliquots of the ultrafiltrate may be assayed to determine C_S^α.

A cause for possible concern in the ultrafiltration techniques (centrifugal or conventional) is the progressive decrease in volume of the reaction mixture as liquid is forced through the membrane to generate the ultrafiltrate. Because the total concentrations of acceptor and ligand in the mixture are therefore constantly changing, an extremely accurate record of the volume and ligand concentration of the ultrafiltrate is needed to determine the values of $\bar{C}_A^\alpha$ and $\bar{C}_S^\alpha$ appropriate to any measured C_S^α (Ford and Winzor, 1981). This method is not recommended for studies of ligand binding to acceptors undergoing self-association, because the relative proportions of monomeric and oligomeric species would then vary with the changing $\bar{C}_A^\alpha$ during collection of ultrafiltrate for measurement of C_S^α. The problem of the changing $\bar{C}_A^\alpha$ may be obviated to a great extent by using the diafiltration reservoir (chamber A in Fig. 2.3a) to supplement the reaction mixture with either buffer or ligand solution as the ultrafiltrate is generated (Blatt *et al.*, 1968; Roy and Miles, 1982). However, because the assumed constancy of the reaction mixture volume is an approximation that is not always justified (Roy and Miles, 1982), errors are introduced into calculations of the amount of buffer/ligand solution introduced from the diafiltration reservoir. The major error in r thus emanates from uncertainty in $\bar{C}_S^\alpha$ rather than $\bar{C}_A^\alpha$.

Adsorption of extremely hydrophobic ligands to the membrane and/or filtration assembly can be a further complicating factor in ultrafiltration, but one that can be overcome by subjecting aliquots of the same reaction mixture to ultrafiltration in the same assembly until successive aliquots yield the same value of C_S^α (Harris and Winzor, 1988a; Wilson *et al.*, 1990). Problems of ligand adsorption are also avoided in diafiltration because of the requirement that filtration be continued until the ligand concentration in the ultrafiltrate matches that in the reservoir. Consequently, the only source of error in r emanates from uncertainty in the value of $\bar{C}_A^\alpha$ because of the minor variation in the volume of the reaction mixture during diafiltration. However, diafiltration does entail the subsequent measurement of $\bar{C}_S^\alpha$, which relies upon the availability of an analytical method that yields the total concentration of ligand in free and bound states. The combination of this limitation with the longer duration of experiments and the requirement for large volumes of ligand solution is undoubtedly responsible for the preferred use of ultrafiltration over diafiltration, despite the theoretical advantages of the latter technique.

2.3. PARTITION EQUILIBRIUM

Although the two previous procedures, equilibrium dialysis and ultrafiltration/diafiltration, are both partition equilibrium methods, a semipermeable membrane is required to create the two phases between which free ligand equilibrates. In this section we focus attention on the characterization of ligand binding by partition studies in systems where the acceptor-containing and acceptor-free phases arise spontaneously. For example, there are many instances where the acceptor (receptor) sites are located on cellular structures, such as the myofibrillar matrix, cytoskeleton, or cell membrane, and where the soluble ligand partitions between the liquid and particulate phases by virtue of its interaction with acceptor sites. In addition, liquid–liquid partition studies have been employed to characterize interactions between an acceptor that is confined to an aqueous phase and a ligand that partitions between the aqueous and a nonaqueous phase. The latter are considered first.

2.3.1. Liquid–Liquid Partition Studies

As an example of the characterization of ligand binding by liquid–liquid partition, we cite classic studies of the interaction between serum albumin and long-chain fatty acids (Goodman, 1958b, Spector *et al.*, 1969). Whereas the amphiphilic nature of the fatty acid allows it to partition between the aqueous (α) and *n*-heptane (β) phases, albumin and albumin–ligand complexes are confined to the aqueous phase (Fig. 2.4a). After attainment of thermodynamic equilibrium between the two phases, the concentrations of free ligand in the aqueous and heptane phases, $(C_S^\alpha)_e$ and $(C_S^\beta)_e$, respectively, are related by the expression

$$(C_S^\beta)_e = K_p(C_S^\alpha)_e \tag{2.7}$$

where K_p is the coefficient describing the partition of free ligand. This coefficient, which may well exhibit dependence on the magnitude of $(C_S^\beta)_e$, is readily determined as the ratio of these two concentrations in experiments with fatty acid alone, the magnitudes of $(C_S^\alpha)_e$ and $(C_S^\beta)_e$ being available directly if radiolabeled fatty acid is used to quantify ligand concentrations (Goodman, 1958a). Corresponding analyses of the two phases in mixtures with a known total concentration of albumin, $\overline{C}_A^\alpha$, yield $(\overline{C}_S^\alpha)_e$ and $(C_S^\beta)_e$ and hence allow evaluation of the binding function as

$$r = [(\overline{C}_S^\alpha)_e - (C_S^\beta)_e/K_p]/\overline{C}_A^\alpha \tag{2.8}$$

for each mixture with $(C_S^\alpha)_e = (C_S^\beta)_e/K_p$ (eq. 2.7), K_p being the value of the partition coefficient appropriate to free ligand concentration $(C_S^\beta)_e$.

There are, of course, also potential situations in which the only measurable

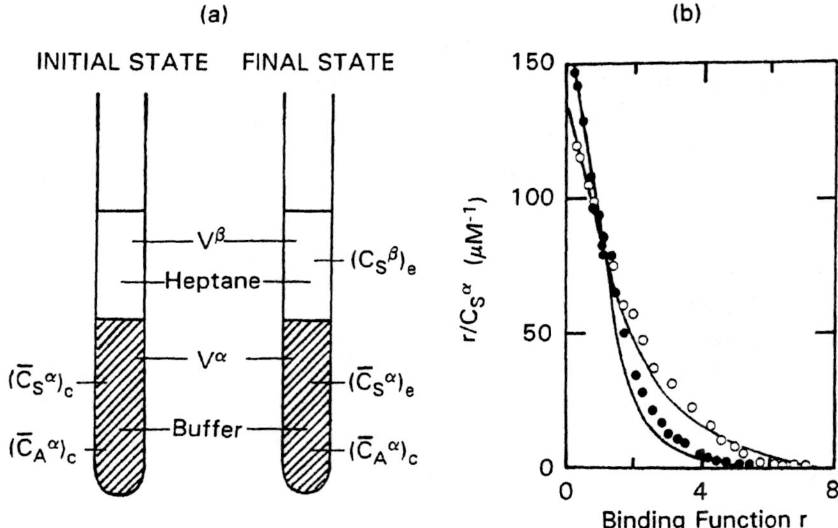

Fig. 2.4. Liquid–liquid partition studies of ligand binding. **(a)** Schematic representation of the initial and final stages of an experiment in order to establish nomenclature. **(b)** Scatchard plots of binding data (Goodman, 1958b) for the interactions of palmitic (○) and stearic (●) acids with bovine serum albumin.

parameter after attainment of equilibrium is the concentration of ligand in the organic phase, $(C_S^\beta)_e$. In accordance with the nomenclature adopted for equilibrium dialysis, we denote total concentrations of acceptor and ligand introduced into the aqueous phase at the commencement of the partition experiment as $(\bar{C}_A^\alpha)_c$ and $(\bar{C}_S^\alpha)_c$. From considerations of mass conservation, it follows that

$$K_p = V^\alpha (C_S^\beta)_e / [V^\alpha (\bar{C}_S^\alpha)_c - V^\beta (C_S^\beta)_e]; \qquad (C_A^\alpha)_c = 0 \tag{2.9}$$

$$r = [V^\alpha (\bar{C}_S^\alpha)_c - (C_S^\beta)_e (V^\beta + V^\alpha / K_p)] / V^\alpha (\bar{C}_A^\alpha)_c \tag{2.10}$$

Comparison of eq. 2.10 with eq. 2.2 reveals the formal identity between expressions for liquid–liquid partition and equilibrium dialysis, particularly on noting that $K_p = 1$ for the latter system due to the aqueous nature of the acceptor-free and acceptor-containing phases. As in equilibrium dialysis, more accurate binding data are obtained under circumstances where the ligand concentrations in both phases at equilibrium can be monitored directly. Scatchard plots (Fig. 2.4b) so obtained for the interactions of palmitic and stearic acids with bovine serum albumin (Goodman, 1958b) are distinctly curvilinear and hence signify that fatty acids bind to more than one class of binding sites on albumin (eq. 1.11).

An obvious limitation of liquid–liquid partition studies for the characterization of ligand binding is the lack of suitable solvents for the second phase, because the requirement for immiscibility with the aqueous solution also diminishes the likelihood of ligand solubility in the nonaqueous phase. Furthermore, even when these requirements are met there is the distinct possibility that acceptor instability on exposure to the organic solvent may well lead to its denaturation and precipitation at the interface between the two phases. An ingenious solution to this problem entails the use of two-phase aqueous polymer systems (Albertsson and Philipson, 1960; Edmond and Ogston, 1968). Although poly(ethylene glycol) and dextran are both water soluble, mixtures thereof separate into two aqueous phases—one richer in dextran and the other richer in poly(ethylene glycol). The fact that both phases are now aqueous clearly overcomes problems in regard to the requirements for ligand partitioning and acceptor stability, but now introduces the complication that acceptor, ligand, and all acceptor–ligand complexes are likely to partition, to various extents, between the two phases. From a re-evaluation (Winzor, 1991) of experimental data for an antigen–antibody system (Elling *et al.*, 1991), it is evident that any attempted quantification requires simplifying assumptions/approximations that may well be difficult to justify. The use of two-phase aqueous polymer systems for characterization of ligand binding is therefore not recommended.

2.3.2. Solid–Liquid Partition Studies

The interaction of a ligand with acceptor sites that are embedded in a solid phase is another situation where the acceptor-free liquid phase provides a direct means of measuring C_S. Mixtures containing known amounts of matrix and ligand are allowed to equilibrate at the temperature of interest until partition equilibrium has been established, at which stage a sample of the supernatant is obtained by centrifugation. The concentration of ligand in the liquid phase is then determined by any appropriate spectral, enzymatic, or radiochemical means. This quantity is unequivocally the equilibrium concentration of free ligand, C_S, for a system with a total ligand concentration, $\overline{C}_S$, that is obtained by dividing the molar amount of ligand added by V_S^*, the volume accessible to ligand. The latter may be obtained by conducting an experiment under conditions where the matrix–ligand interaction is effectively eliminated either by the presence of a competitive inhibitor or by suppression of k_{AS} to a negligible magnitude (through change in pH, temperature, or ionic strength). Because the concentration of acceptor sites ($p\overline{C}_A$) is of unknown magnitude, this parameter also needs to be evaluated from the $[(\overline{C}_S - C_S), C_S]$ results as the concentration of ligand bound per unit weight of acceptor in the limit of infinite free ligand concentration. One procedure for so doing is illustrated in Figure 2.5, which presents in double-reciprocal format the binding of aldolase to muscle myofibrillar suspensions (Kuter *et al.*, 1983; Harris and Winzor, 1985). The reciprocal of the ordinate intercept signifies a capacity of 78 nmol aldolase per gram of myofibrillar protein. No attempt is made to quantify the binding affinity at this stage, because the curvilinear form of Figure 2.5 does not reflect matrix-site heterogeneity (eq. 1.11) but rather multivalency of the enzymatic ligand—a problem addressed in Chapter 7.

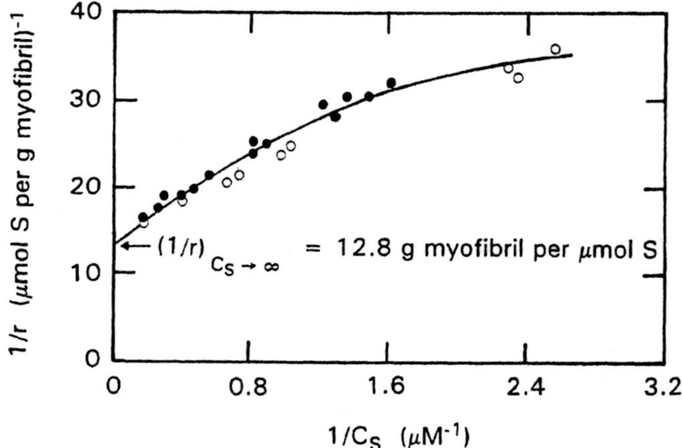

Fig. 2.5 Evaluation of the binding capacity of rabbit skeletal muscle myofibrils for aldolase from partition equilibrium studies. Solid symbols refer to results obtained by Harris and Winzor (1985) and open symbols to results of Kuter *et al.* (1983).

An obvious disadvantage of such partition experiments is the need for precise determination of the amount of matrix introduced into each reaction mixture—a potential source of uncertainty in instances where aliquots are being taken from a concentrated slurry of cellular matrix. In some instances this problem may be overcome by using a recycling partition technique (Ford and Winzor, 1981; Hogg *et al.*, 1991) in which the liquid phase of a stirred slurry of matrix and ligand is monitored spectrophotometrically by means of a flow cell placed

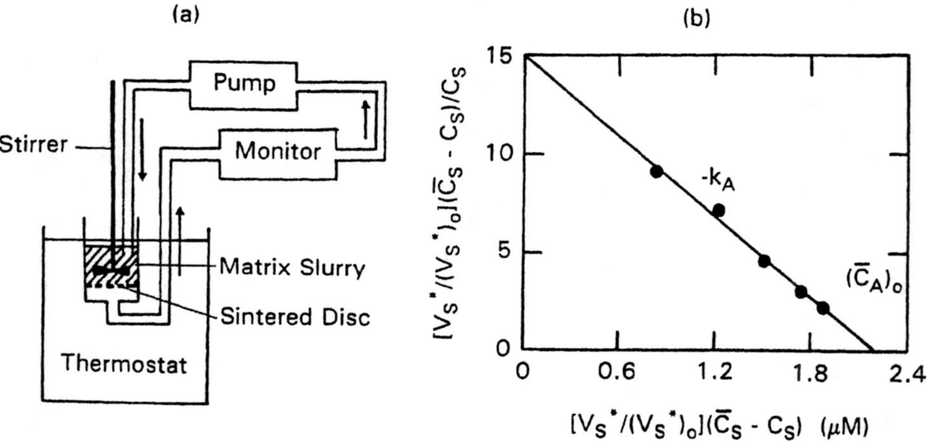

Fig. 2.6. Studies of ligand binding by the recycling partition technique. **(a)** Schematic representation of the experimental system. **(b)** Scatchard plot of results so obtained (Hogg *et al.*, 1991) for the interaction of antithrombin III with heparin immobilized on Sepharose.

in the line returning the liquid phase to the slurry (Fig. 2.6a). This procedure has the advantage that a number of partition experiments may be performed with the same sample of matrix by making successive additions of concentrated ligand solution to the slurry and determining and corresponding value of C_S after each addition.

Although the recycling partition technique avoids any ambiguity about constancy of the amount of matrix, the successive measurements of C_S after each ligand addition refer to a system with ever-decreasing $\overline{C}_A$ because of the increased volume of the mixture. This effect of dilution may be taken into account by multiplying each $(\overline{C}_S - C_S)$ result by $V_S^*/(V_S^*)_o$, where $V_S^* = (V_S^*)_o + \Sigma_1^i \delta V$, the volume accessible to ligand being increased from its original value, $(V_S^*)_o$, by a volume increment δV after each addition of ligand. The value of $p\overline{C}_A$ so obtained then refers to the system with matrix in the original accessible volume, $(V_S^*)_o$. A Scatchard representation of results obtained by the recycling partition technique for the interaction of antithrombin with heparin immobilized on Sepharose (Hogg *et al.*, 1991) is presented in Figure 2.6b.

2.4. GEL CHROMATOGRAPHY

Binding data are obtained readily and rapidly by gel chromatography, sometimes referred to as *exclusion chromatography* or *molecular sieve chromatography*. The only proviso to the use of this technique is that acceptor (A) and all complexes (AS_i) co-migrate (Nichol and Winzor, 1964; Nichol *et al.*, 1971a). This requirement that acceptor and acceptor–ligand complexes possess identical elution volumes ($V_A = V_{AS_i}$) is usually accomplished by selecting a gel phase that excludes acceptor, whereupon it and all complexes are confined to the liquid phase surrounding the gel beads of the column and hence are eluted in the void volume (V_o). Thus gel chromatography on matrices such as Sephadex G-25 and Bio-Gel P-2, which exclude acceptors with $M_A > 2{,}000$, is suitable for studying protein–drug (Cooper and Wood, 1968), enzyme–inhibitor (Tellam *et al.*, 1979) and enzyme–coenzyme (Ward and Winzor, 1983) interactions. Furthermore, the size of the ligand can be increased markedly by selecting a chromatographic matrix with larger pores. For example, a column of porous glass beads with a pore diameter of 170 nm has been used (Hogg and Winzor, 1985) to quantify the interaction of concanavalin A ($M_S = 50{,}000$) with a dextran acceptor ($M_A = 2 \times 10^6$). Consequently, gel chromatography is more versatile than equilibrium dialysis for studies of ligand binding in that the only requirement is a large difference between the sizes of acceptor and ligand, there being much less restriction on the maximal size of the smaller reactant (ligand).

2.4.1. Frontal Gel Chromatographic Studies

Like equilibrium dialysis and ultrafiltration, frontal gel chromatography (Nichol and Winzor, 1964) has the potential for direct measurement of the equilibrium concentration of free ligand (C_S^α) in a mixture with $\overline{C}_S^\alpha$ and $\overline{C}_A^\alpha$ as the respective total concentrations of ligand and acceptor. To create a plateau region with defined composition ($\overline{C}_A^\alpha$, $\overline{C}_S^\alpha$) in the elution profile, a mixture of acceptor and ligand in buffer is applied to the packed gel chromatographic column until the emerging effluent has the same composition as the solution being applied. This entails the addition of a volume of mixture at least equal to V_S, the elution volume of ligand. Elution of the column with buffer then generates a trailing elution profile such as that shown in Figure 2.7a for a mixture of sulfamethoxypyridazine and albumin (Cooper and Wood, 1968) that had been subjected to prior equilibrium dialysis to determine the magnitude of C_S^α. The elution profile comprises two boundaries: a reaction boundary at V_A that separates the plateau of original composition from a plateau of pure ligand and a boundary at V_S corresponding to elution of pure ligand. Quantitative considerations (Nichol and Winzor, 1964; Nichol *et al.*, 1971a) have established that the concentration of the ligand plateau, C_S^β, is equal to the free ligand concentration in the applied mixture, C_S^α, this being a prediction borne out in Figure 2.7a by the identity of C_S^β and the concentration of ligand in the diffusate (C_S^α) from the prior equilibrium dialysis step. Clearly, the dialysis step is redundant inasmuch as the frontal gel chromatography experiment has the ca-

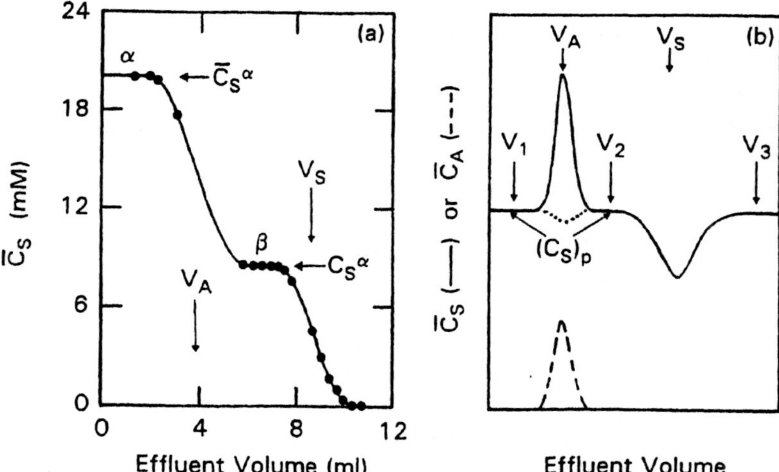

Fig. 2.7. Studies of ligand binding by gel chromatography. **(a)** Trailing elution profiles, deduced from Cooper and Wood (1968), for a mixture of serum albumin (14 mM) and sulfamethoxypyridine (20 mM) on a column (0.5 × 30 cm) of Sephadex G-25: C_S^α denotes the estimate of the free drug concentration obtained by equilibrium dialysis. **(b)** Schematic Hummel-Dreyer elution profiles for ligand (solid line) and acceptor (dashed line) after application of a small zone of acceptor–ligand mixture to a gel column pre-equilibrated with the same concentration of ligand, $(C_S)_p$, as that, $(\overline{C}_S^\alpha)_a$, in the applied mixture. The dotted line indicates the variation of free ligand concentration within the acceptor zone in instances where acceptor migrates faster than acceptor-ligand complex(es) (Cann *et al.*, 1989).

pacity to provide the same information (Nichol and Winzor, 1964; Nichol *et al.*, 1971b).

The rapidity with which equilibrium binding results may be acquired is a decided asset of the gel chromatographic technique. For example, generation of the plateau of original composition (α-phase) and development of the elution profile shown in Figure 2.7a was completed in 2 hours (Cooper and Wood, 1968); and an even shorter time period (40 minutes) sufficed to evaluate C_S^β for a lectin–dextran system (Hogg and Winzor, 1985). Further curtailment of the time factor, and also of the amount of acceptor–ligand mixture required, should be achievable fairly readily by resort to the smaller columns and more sophisticated effluent scanners currently being used in high performance liquid chromatography.

2.4.2. The Hummel and Dreyer Procedure

A disadvantage of the above technique for studies of ligand binding is the relatively large amount of acceptor–ligand mixture (at least one column volume) that is required to generate the plateau of original composition ($\overline{C}_A^\alpha$, $\overline{C}_S^\alpha$). An alternative procedure, devised by Hummel and Dreyer (1962), entails the pre-equilibration of a Sephadex G-25 or Bio-Gel P-2 column with a known concentration of ligand, $(C_S)_p$, prior to application of a small zone (volume V_a) of acceptor with concentration $(\overline{C}_A)_a$, dissolved in the ligand solution used for pre-equilibration. Elution of the column is then continued with more of the pre-equilibrating ligand solution. In the resultant elution profile, depicted schematically as the solid line in Figure 2.7b, the increased total ligand concentration ($\overline{C}_S$) coincident with elution of acceptor at V_A ($= V_{AS_i}$) reflects the binding of ligand. Consequently, the amount of ligand bound may be calculated by trapezoidal integration to obtain the area of the peak, provided that a method is available for the determination of total ligand concentration in the presence of acceptor. Alternatively, because complex formation has been achieved at the expense of the pre-equilibrating liquid concentration, the elution profile necessarily exhibits a trough at the elution volume of ligand (V_S), the area of which must also correspond to the amount of complexed ligand. The binding function may therefore be determined as either

$$r = \sum_{V_1}^{V_2} [\overline{C}_S - (C_S)_p]\, \Delta V/(\overline{C}_A)_a \tag{2.11a}$$

or

$$r = \sum_{V_2}^{V_3} [(C_S)_p - \overline{C}_S]\, \Delta V/(\overline{C}_A)_a \tag{2.11b}$$

depending on whether the area of the peak or the trough is being measured.

Greater economy with regard to the amount of acceptor, though not of ligand (Cooper and Wood, 1968), may well be an advantage of this technique. However, the required trapezoidal integration places stringent demands not only on the accuracy with which ligand concentrations are being measured but also on the accuracy with which the volume scale of the elution profile is being defined. In experiments with discontinuous analysis of the column effluent (Fairclough and Fruton, 1966), the size of fractions should therefore really be determined by weight (Winzor and Scheraga, 1963). As noted by Colman (1972), the need for accurate volume determination is eliminated in situations where separate assay procedures allow the elution profile to be defined in terms of the total concentrations of both acceptor and ligand. Under these circumstances the binding function is obtained by substituting the pair of concentrations, $(\overline{C}_S)_V$ and $(\overline{C}_A)_V$, for any given volume V within the acceptor zone into the expression

$$r = [(\overline{C}_S)_V - (C_S)_p]/(\overline{C}_A)_V; \qquad V_1 < V < V_2 \tag{2.12}$$

Furthermore, because eq. 2.12 may be applied to several pairs of $[(\overline{C}_S)_V, (\overline{C}_A)_V]$ values, the binding function for a given free ligand concentration, $(C_S)_p$, may be assigned a mean value and an estimate of its inherent uncertainty.

We note that an exact counterpart of a Hummel-Dreyer gel chromatographic profile has been obtained by counterion electrophoresis of a negatively charged calcium-binding protein in the presence of Ca^{2+} ions (Ueng and Bronner, 1979). Consequently, despite the use of electrophoresis rather than gel chromatography to generate the uniform concentration of free ligand throughout the pattern, pairs of $[\overline{C}_S, \overline{C}_A]$ values from within the acceptor zone were also amenable to quantitative interpretation in terms of eq. 2.12, with $(C_S)_p$ identified as the uniform concentration of free calcium ion.

2.4.3. Limitations of the Gel Chromatographic Procedures

Mention has already been made of the requirement in frontal gel chromatography that the acceptor and all complexes co-migrate for C_S^β to be identified as the free ligand concentration in the applied mixture. Nonconformity with this requirement does not affect the general form of the the elution profile (Fig. 2.7a), but the concentration of the pure ligand plateau, C_S^β, is no longer C_S^α (Gilbert and Jenkins, 1959; Nichol and Winzor, 1964). More complicated analyses are therefore required, even in instances where the interaction involves 1:1 stoichiometry (Nichol and Winzor, 1964; Nichol *et al.*, 1967b; Gilbert and Kellett, 1971; Cann and Winzor, 1987).

The validity of the Hummel and Dreyer procedure is also subject to the same restrictive requirement, which poses no problems in studies of protein–ligand interactions by gel chromatography on Sephadex G-25 or Bio-Gel P-2. However, use of the Hummel and Dreyer method for studies of metal–nucleotide interactions on Sephadex G-10 (Colman, 1972) represents an invalid application of the technique because the metal-nucleotide complex exhibits an elution volume (V_{AS}) intermediate between those of nucleotide, V_A, and metal ion, V_S (Cann *et al.*, 1989). Numerical simulation of the Hummel-Dreyer profiles for the ATP–Mg^{2+} system on Sephadex G-10 has shown (Cann *et al.*, 1989) that the free ligand concentration within the acceptor zone decreases below $(C_S)_p$ (dotted line in Fig. 2.7b) for this combination of species elution volumes ($V_A < V_{AS} < V_S$), in which case determination of the binding function from eq. 2.11 or 2.12 underestimates the true value. On the other hand, faster migration of the complex than of either reactant ($V_C < V_A < V_S$) leads to a free ligand concentration greater than $(C_S)_p$ within the acceptor zone and hence overestimation of r (Cann *et al.*, 1989). An alternative gel chromatographic procedure for characterizing metal–nucleotide interactions is discussed in Chapter 4, which considers evaluation of the free ligand concentration by methods that do not rely upon generation of an acceptor-free ligand phase.

3

SPECTRAL STUDIES OF LIGAND BINDING

Having considered methods for evaluation of the binding function by measuring the free ligand concentration, C_S, in mixtures with total concentrations $\overline{C}_A$ and $\overline{C}_S$ of acceptor and ligand, respectively, we now turn to spectral methods for the measurement of $\overline{C}_S - C_S$, the difference between total and free ligand concentrations in such mixtures.

3.1. GENERAL CONSIDERATIONS

The association of a ligand with an acceptor to form a complex is often accompanied by perturbation of the spectral characteristics of one of the reactants. As an example, we cite the perturbation of the heme spectrum of microsomal cytochrome P-450 by ligands such as hexobarbital (Estabrook *et al.*, 1972). The spectral change (Fig. 3.1a) not only tells us something about the groups in the molecules that may be affected by the association phenomenon but also provides information for quantifying the binding equilibrium itself. In some cases the sensitivity of the spectroscopic technique makes it the method of choice for monitoring the binding, particularly in instances where limited amounts of ligand or acceptor are available. Spectroscopic titrations usually involve the measurement of a spectral parameter as a function of $\overline{C}_S$, the total concentration of ligand added (Fig. 3.1b). Because the spectral parameter is related to the concentration of complex, the free concentration of ligand is determined as the difference between $\overline{C}_S$ and $\overline{C}_S - C_S$, the concentration of ligand involved in complex formation with acceptor. In this section we present initially a general treatment for the evaluation of binding data from titrations in which the spectral signal arises from the acceptor; we then continue with

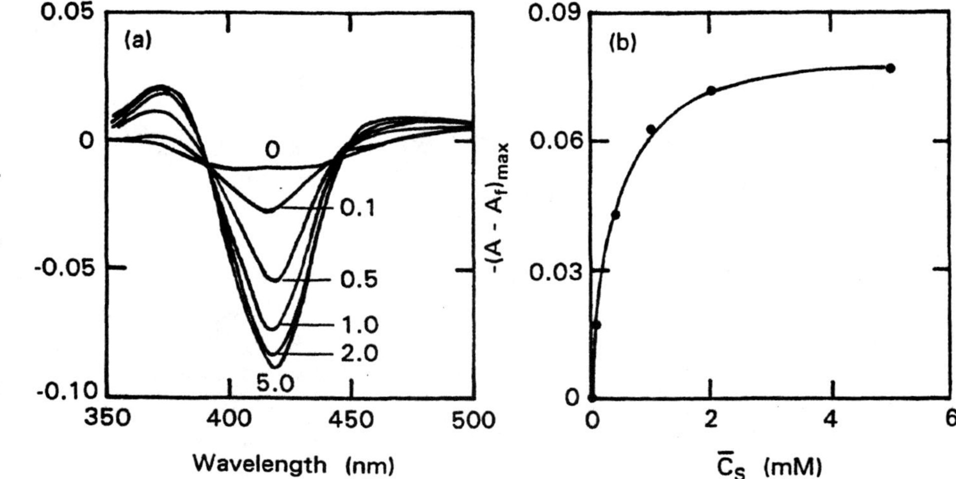

Fig. 3.1. The interaction of hexobarbital with microsomal cytochrome P-450 as an example of a binding titration in which the spectroscopic signal arises from the acceptor. **(a)** Difference spectra showing the perturbation of the heme spectrum of the microsomal cytochrome P-450 (1 mg protein/ml) in the 400–450 nm region by the indicated concentrations (mM) of hexobarbital. **(b)** Plot of the maximal perturbation as a function of hexobarbital concentration, $\overline{C}_S$. Experimental data are taken from Estabrook *et al.* (1972).

the corresponding consideration of titrations in which the ligand is the source of the spectral signal.

3.1.1. Purity and Stability of Preparations

Spectroscopic techniques demand spectroscopic purity. That is, the signal must arise from a single species. The presence of impurities that contribute to the spectroscopic signal must therefore be avoided. Small levels of background signal such as solvent fluorescence and scattered excitation light in fluorescence titrations may be tolerated provided that appropriate blank corrections are made. These signals are inert in that they do not interfere with the reaction under study. Difficulties arise if the impurity, whether contributing to the spectroscopic signal or not, interferes with the binding process by binding to the ligand or by competing with the ligand for sites on the acceptor. Similarly, it is desirable to avoid situations in which the spectroscopic signal arises from several species that participate, to varying degrees, in the binding reaction. Because quantitative investigations of binding affinity and interaction stoichiometry require reactant preparations of high purity, much time can be saved by a thorough investigation of this question prior to commencing spectral binding experiments. It is necessary to have an accurate estimate of the total concentration of ligand, since the free ligand concentration is inferred as the difference between $\overline{C}_S$ and $\overline{C}_S - C_S$. Similarly, an accurate value of the total acceptor concentration, $\overline{C}_A$, is required for evaluation of the binding function (eq. 1.3). For the determination of $\overline{C}_A$ it is unwise to rely upon a colorimetric assay (Lowry or dye-binding method) for determining the concentration of a protein acceptor because of the marked dependence of the color yields upon aromatic content. It is better to rely upon an extinction coefficient that has been based on an amino acid analysis, Kjeldahl nitrogen estimation, or a differential refractive index measurement.

Despite the above emphasis on the need for reactant purity in quantitative studies of acceptor–ligand interactions by spectroscopic titration, it should be kept in mind that the aim of binding experiments in biology is to characterize *in vivo* phenomena. Consequently, the biochemist aims to work toward *in vivo* complexity, and in so doing may ultimately have to sacrifice experimental rigor. Some experimental techniques with high resolving power have the potential to provide useful binding information by identifying uniquely a signal in extremely complicated biological systems. Noteworthy among these are the magnetic resonance techniques (NMR, ESR) and, to a lesser extent, fluorescence assays. For example, the ^{31}P NMR signal from the γ-phosphate group of ATP within cells of a cellular suspension provides a measure of proton binding by ATP and therefore an estimate of the intracellular pH (Bailey *et al.*, 1981). Inasmuch as the spectroscopic signal must still be attributable to the component of interest in these complex systems, the requirement for spectral purity remains.

3.1.2. Time Dependence of Signals

Binding reactions usually take place on the millisecond time scale or faster, and hence time dependence of experimental signals is not usually observed unless rapid reaction (temperature-jump, stopped-flow) techniques are used. However, there are circumstances in which the spectroscopic signal may develop slowly because of factors such as a slow, time-dependent conformational change within the acceptor, viscosity-limited diffusion of ligand to the acceptor, or indeed, retarded access of ligand to acceptor across a permeable barrier. Thus, binding to cell surface receptors is frequently time dependent, possibly due to the necessity of ligand diffusion through a glycocalix barrier or to changing receptor accessibility as the result of cell membrane dynamics. It is therefore important to check for time dependence to establish that measurements are taken under equilibrium conditions.

3.1.3. Limitations of Spectroscopic Binding Titrations

Certain experimental considerations limit the design of spectroscopic binding assays, two being of particular concern.

1. The change in the spectroscopic signal must be large enough to be measured accurately. For this purpose it is preferable to employ difference spec-

troscopy in which the reference and sample cuvets contain the chromophore (fluorophore) at the same concentration. Such procedures not only enhance the intrinsic sensitivity of spectroscopy but also provide data in readily usable form. For weak interactions ($k_{AS} = 10^3 - 10^4\ M^{-1}$) the concentrations of ligand and acceptor are relatively high, and hence measurement of their concentrations is rarely a problem; but unfortunately the amounts of acceptor available are limited for many biological systems. On the other hand, only small amounts of acceptor and ligand are required for the characterization of high-affinity interactions ($k_{AS} = 10^8 - 10^{10}\ M^{-1}$); but the sensitivity of detection may then become limiting. Although radiochemical-based assays have traditionally been used in such cases, fluorescent and enzyme-linked assays are now starting to approach the sensitivity of their radiochemical counterparts.

2. Where concentrations of the ligand or acceptor are less than 10^{-6} M, a significant fraction of either or both species may become adsorbed to the walls of the cuvets. Clearly, the degree of surface adsorption needs to be checked in control experiments. Surface adsorption can sometimes be minimized by pretreatment of the glass surfaces with dichlorodimethylsilane (Ferdinand, 1964; Cecil and Robinson, 1975) or by inclusion of an inert solute in all solutions.

3.2. A SPECTROSCOPIC SIGNAL ARISING FROM THE ACCEPTOR

The results of a typical binding titration in which the spectroscopic signal arises from the acceptor have already been shown in Figure 3.1. Provided that each acceptor site occupied causes an equivalent change in the spectroscopic signal, the fraction of binding sites occupied, f_a, is given by the fractional change in the spectroscopic parameter (y). Thus,

$$f_a = (y - y_f)/(y_b - y_f) \tag{3.1}$$

where y_b and y_f are the signals from the acceptor when sites are fully occupied and unoccupied, respectively. At this stage the concentration of bound ligand cannot be determined unless the number of binding sites (stoichiometry of the reaction) is known. This can be found from a separate experiment.

3.2.1. Determining the Stoichiometry

This titration is carried out under conditions where the acceptor concentration is much higher than the dissociation constant (reciprocal of the association constant) of the binding reaction. Titration data can be transformed into a plot of fractional saturation against the ratio of total ligand added to total acceptor, as in Figure 3.2. Reaction stoichiometry is then inferred by extrapolating the two linear segments of the graph to their point of intersection, which represents the endpoint of the titration. Curve a of Figure 3.2 emphasizes the point that

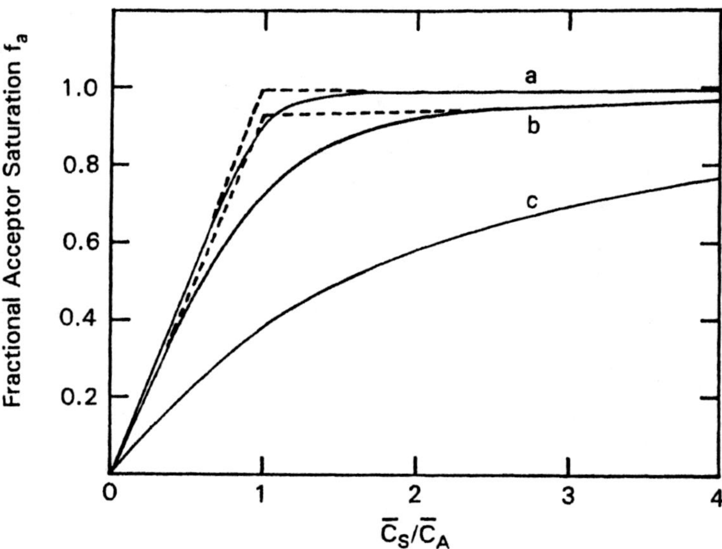

Fig. 3.2. Evaluation of the stoichiometry of an acceptor–ligand interaction by stoichiometric titration. Curves have been constructed (eqs. 1.9a, 3.1–3.3) on the basis of a single acceptor site with a binding constant (k_{AS}) of 10^6 M^{-1} and values of **(a)** 10^{-4} M, **(b)** 10^{-5} M, and **(c)** 10^{-6} M for the fixed total concentration of acceptor ($\bar{C}_A$).

this extrapolation is made readily when the concentration of acceptor greatly exceeds the dissociation constant for the interaction ($k_{AS}\bar{C}_A = 100$ in this instance). In practice, the condition $k_{AS}\bar{C}_A = 10$ is usually considered sufficient to ensure such stoichiometric titration of acceptor, *i.e.*, to ensure the essential absence of any free ligand until all sites on the acceptor have been titrated. However, from curve b of Figure 3.2 this condition is clearly the lower limit for extrapolation of two linear segments to obtain the stoichiometry. High-affinity binding is thus an advantage from the viewpoint of acceptor concentration required to achieve the stoichiometric binding conditions that give rise to the titration curve comprising two linear segments. Once the stoichiometry has been determined it is possible to evaluate the binding constant from a titration under conditions where there are varying proportions of free and bound ligand present during the binding titration (curve c of Fig. 3.2).

3.2.2. The Equilibrium Titration

This titration is carried out under conditions where the acceptor concentration is close to the dissociation constant for the binding equilibrium, in which case the titration curve assumes a more rounded form (Fig. 3.2 curve c). The concentrations of free and bound ligand can be found from the values of the fractional saturation and the number of binding sites (stoichiometry). For this purpose the fractional saturation is converted to the binding function, r,

$$r = f_a p \tag{3.2}$$

whereupon the concentration of free ligand becomes

$$C_S = \overline{C}_S - r\overline{C}_A \tag{3.3}$$

The data can now be plotted as r *versus* C_S or one of its linear transforms and the association constant determined by any of the analyses discussed in Chapter 1. In that regard it should be noted that an internal check is provided in that the number of binding sites (p) deduced from the analysis should agree with that from the stoichiometric titration.

In summary, two separate titrations are usually performed to determine binding parameters in spectrophotometric studies where the signal arises from the acceptor (Uehara *et al.*, 1978; Heyduk and Lee, 1990). The first titration uses a relatively high acceptor concentration to establish the stoichiometry of the reaction, whereas the second titration is carried out at a lower acceptor concentration that is conducive to the existence of an equilibrium between free and bound ligand. Examples of the two titrations are presented in Figure 3.3a,b for the interaction of *Escherichia coli* cAMP receptor protein with the primary site on the *lac* promoter (Heyduk and Lee, 1990), while the suggested analysis of the resulting [r, C_S] data is shown in Figure 3.3c. For this system the interaction was detected by its effect on the polarization (anisotropy) of fluorescence of fluorescently labeled DNA. Accordingly the ordinate axes for the two titrations are expressed in terms of fluorescence anisotropy rather than fluorescence intensity. Furthermore, the magnitude of k_{AS} (6.0×10^7 M^{-1}) obtained from Figure 3.3c is an apparent value dependent upon the concentration of cyclic nucleotide because reaction of the receptor protein with cAMP is a prerequisite for its interaction with the *lac* promoter.

3.2.3. Special Cases

Occasionally the number of binding sites on the acceptor is known with certainty. For example, X-ray diffraction studies have shown that hemoglobin has four binding sites for oxygen and that immunoglobulin G has two sites for antigen. Enzymes often have a single site for substrate and/or substrate analog. In such cases the equilibrium titration alone suffices to establish the thermodynamic parameters of binding or the structural requirements of a set of ligand analogs that direct the specificity of binding. However, in view of the assumptions involved in spectroscopic binding assays, it is advisable to confirm data by other methods. Because of its rigorous thermodynamic basis, equilibrium dialysis is often regarded as a gold standard against which all other methods can be compared.

The binding of oxygen to hemoglobin deserves special mention in view of its historical role in the development of allosteric theory. This interaction has pronounced effects on the absorption spectrum of the heme prosthetic group,

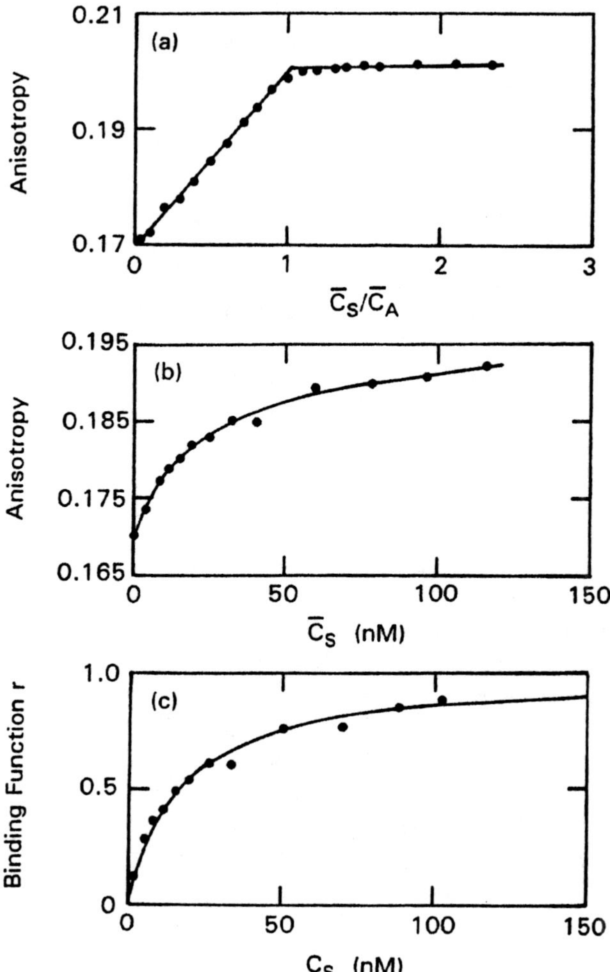

Fig. 3.3. Binding of *Escherichia coli* cAMP receptor protein to a 32-base-pair DNA fragment of the *lac* promoter, as measured by fluorescence polarization (anisotropy). **(a)** Stoichiometric titration of the fluorescently labeled DNA fragment ($\bar{C}_A$ = 98.5 nM) with receptor protein in the presence of 0.2 mM *c*AMP. **(b)** Equilibrium titration ($\bar{C}_A$ = 11.1 nM) in the presence of 0.5 μM *c*AMP. **(c)** Secondary plot of the equilibrium titration data as a binding curve, the analysis of which in terms of eq. 1.9a with p = 1 leads to an apparent association constant of 6.0×10^7 M^{-1}. Experimental data are taken from Heyduk and Lee (1990).

the most substantial changes occurring at wavelengths of 400–405 and 542–576 nm, any of which may be used to determine the fractional saturation (f_a) via eq. 3.1. The partial pressure of oxygen in the atmosphere above the hemoglobin solution is proportional to that in free solution, and the abscissa of the binding curve is therefore often expressed as the partial pressure of oxygen in units of mm Hg or, in keeping with SI practice, kilopascals (kP). To adjust the concentration of oxygen in equilibrium with the hemoglobin, the partial pressure of oxygen in the air above the solution is changed by connecting the spectrophotometer cuvet to a pressure vessel known as a *tonometer* (Allen *et al.*, 1950). Because the concentration of oxygen is now frequently measured directly with an oxygen electrode (Imai *et al.*, 1970), it can be expressed either as the partial pressure of oxygen in equilibrium with the hemoglobin solution or as the concentration of oxygen free in solution. Edsall and Gutfreund (1983) have pointed out that although this choice does not affect the relative ratios of the equilibrium constants for the four binding steps, it does affect their magnitudes and dimensions. This manner of description is not important when successive equilibrium constants are being compared, but does become important in calculations of reaction enthalpies via the van't Hoff relationship (eq. 1.1). If binding equilibria are described in terms of partial pressures, the calculated enthalpy change ($\Delta H°$) is the sum of the enthalpies for the binding reaction and the equilibrium between oxygen in the atmospheric and solution phases. On the other hand, the value of $\Delta H°$ obtained on the basis of equilibrium constants expressed in terms of the concentration of free oxygen in the solution phase refers only to the binding reaction.

3.3. A SPECTRAL SIGNAL ARISING FROM THE LIGAND

3.3.1. Determining Spectral Parameters of Free and Bound Ligand

Certain spectral properties of the ligand may be changed on binding to the acceptor. Its extinction coefficient may change, its absorption maximum may change, or its fluorescence may be quenched or enhanced. The first step in using such changes for the characterization of ligand binding is to determine the spectral properties of the ligand in free and bound states. Whereas the characteristics of the free ligand can be determined by direct measurement, those of the bound ligand can be found by titrating a fixed concentration of ligand with acceptor. After correction for dilution factors, the spectral data are usually plotted in double-reciprocal format ($1/y$ *vs.* $1/\overline{C}_A$) to obtain the spectral parameter for bound ligand as the reciprocal of the ordinate intercept. Although it is frequently assumed that this extrapolation to infinite acceptor concentration ($1/\overline{C}_A \rightarrow 0$) is linear, examination of the theory shows that this is not the case. The curvilinearity of such plots is evident not only in simulated data (Fig. 3.4a) but also in the experimental plot (Fig. 3.4b) for the binding of 12-(9-anthroyloxy)-stearic acid to egg phosphatidylcholine liposomes (Haigh and Sawyer, 1978). Clearly, nonlinear regression analysis is required to characterize the fluorescence intensity of the bound ligand.

The spectroscopic signals that are obtained for the free (y_f) and bound (y_b) ligand by the above procedure are then usually assumed to pertain to all ligand concentrations, but this need not necessarily be the case. For example, concentration quenching due to the inner filter effect may occur at high concentra-

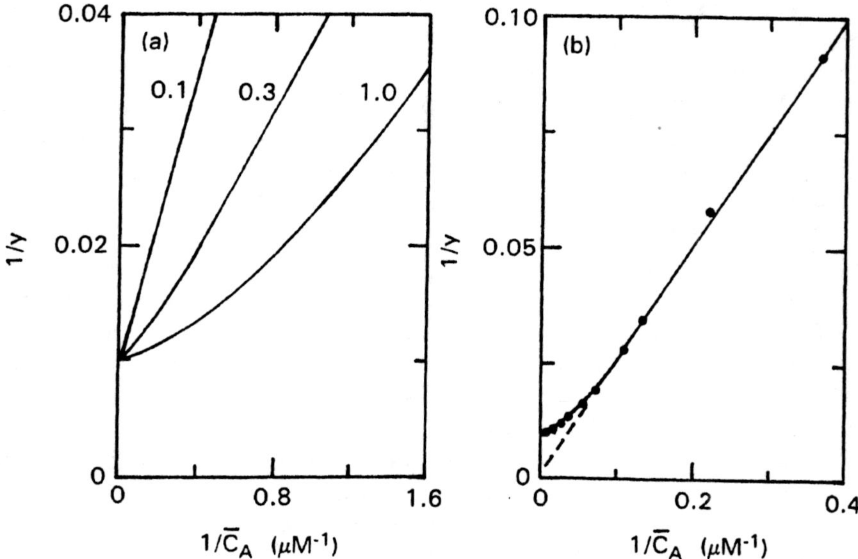

Fig. 3.4. Determination of the spectral parameter for bound ligand (y_b) as the ordinate intercept of a double-reciprocal plot. **(a)** Simulated data for systems with $y_b = 100$, $y_f = 0$, $\overline{C}_S = 3\ \mu M$, $p = 2$, and the indicated values (μM^{-1}) of k_{AS}. **(b)** Experimental results, normalized to give $y_b = 100$, for the binding of 12-(9-anthroyloxy)-stearic acid ($C_S = 3.1\ \mu M$) to egg phosphatidylcholine liposomes. [Adapted from Haigh and Sawyer (1978).]

tions of a fluorescent ligand (Lacowicz, 1983); and deviations from the Beer-Lambert law may well occur in concentrated solutions of a chromophore.

3.3.2. The Equilibrium Titration

A second titration is now carried out under equilibrium conditions such that a constant concentration of acceptor ($\overline{C}_A$) is titrated with increasing amounts of ligand. At any given ligand concentration, $\overline{C}_S$, the fraction of ligand bound (f_s) may be calculated from the measured signal (y) and the respective values for bound and free ligand (y_b, y_f) via the relationship

$$f_s = (y - y_f)/(y_b - y_f) \tag{3.4}$$

whereupon the expressions for the binding function (r) and free ligand concentration (C_S) become

$$r = f_s\overline{C}_S/\overline{C}_A \tag{3.5}$$

$$C_S = \overline{C}_S - f_s\overline{C}_S \tag{3.6}$$

These calculations are repeated for each concentration of ligand added to allow the construction of a complete binding curve (r *vs.* C_S), the analysis being greatly simplified if either the bound or the free ligand yields zero spectral signal (*i.e.*, if $y_b = 0$ or $y_f = 0$).

There are assumptions inherent in the above analysis that depend on the ability of the spectroscopic method to differentiate between the ligand and the acceptor. If part of the signal arises from the acceptor as well as the ligand, suitable controls must be carried out and corrections made accordingly. Such a situation in fluorescence titrations has been discussed by Pesce *et al.* (1971).

3.4. SIMULTANEOUS EVALUATION OF BOTH BINDING PARAMETERS

Each of the above spectral procedures for characterizing ligand binding has entailed two separate titrations. In instances where the spectral signal arises from acceptor, a preliminary titration of acceptor with ligand under conditions of essentially stoichiometric binding is required to establish the number of ligand-binding sites on the acceptor. Similarly, in instances where the spectral signal is a property of the ligand, a solution of the latter must first be titrated with acceptor to obtain the spectral parameter for bound ligand. We now consider two procedures that allow the simultaneous measurement of the stoichiometry (p) and binding constant (k_{AS}) from a titration under equilibrium conditions.

3.4.1. The Dilution Titration

This method, introduced by Anderson and Weber (1965), can be applied to situations in which either the acceptor or the ligand provides the spectroscopic signal. The titration is started with concentrations of acceptor and ligand conducive to stoichiometric binding ($k_{AS}\overline{C}_{AS} \approx 10$; $\overline{C}_S$ slightly greater than the concentration of acceptor sites, $p\overline{C}_A$). Under these conditions the fractional saturation of acceptor sites is about 0.9, and the spectral signal should approximate that of bound ligand or saturated acceptor, whichever is being monitored. The concentration of the mixture is then decreased 100-fold by addition of buffer in 15–30 steps, by which stage the fractional saturation is about 0.1. At each stage of the titration the fractional saturation of the acceptor or the fraction of ligand bound is determined and the binding function calculated for a unique concentration of free ligand (eqs. 3.1–3.3 or 3.4–3.6, as appropriate).

A potential source of error in the dilution titration is the assumed absence of acceptor self-association, a phenomenon that could invalidate the analysis because of the dissociation of acceptor by dilution. Only in the event that monomeric and polymeric acceptor species were to bind ligand with equal affinity would such a procedure yield a valid binding curve (Nichol *et al.*, 1967a, 1972a). Special precautions are also required in instances where the ligand undergoes self-association (Nichol *et al.*, 1969; Sculley *et al.*, 1981), a situation pertinent to the binding of dimeric repressor proteins to DNA operator sites.

3.4.2. Iterative Determination of Both Binding Parameters

A second spectral technique that is preferable from the above viewpoint but more demanding in its requirement for manipulation of the experimental measurements has been devised to allow elucidation of the stoichiometry as well as the equilibrium constant from a single equilibrium titration of acceptor with ligand in instances where acceptor sites are equivalent and independent (Holbrook, 1972; Holbrook *et al.*, 1972). Results such as those presented in Figure 3.5 from studies of the interaction between chloroquine and porphyrin by difference spectroscopy (Shanley *et al.*, 1985) are first analyzed on the basis of $\overline{C}_S$ as an estimate of C_S to obtain an initial value of the spectral property for fully complexed acceptor, $(A_b - A_f)$ in Figure 3.5a and hence first estimates of f_a (eq. 3.1) for each $\overline{C}_S$. Advantage is then taken of the fact that $C_S = \overline{C}_S - f_a p\overline{C}_A$ (eq. 3.3), whereupon it follows from the law of mass action that

$$k_{AS} = f_a p\overline{C}_A / [(1 - f_a) p\overline{C}_A (\overline{C}_S - f_a p\overline{C}_A)] \qquad (3.7a)$$

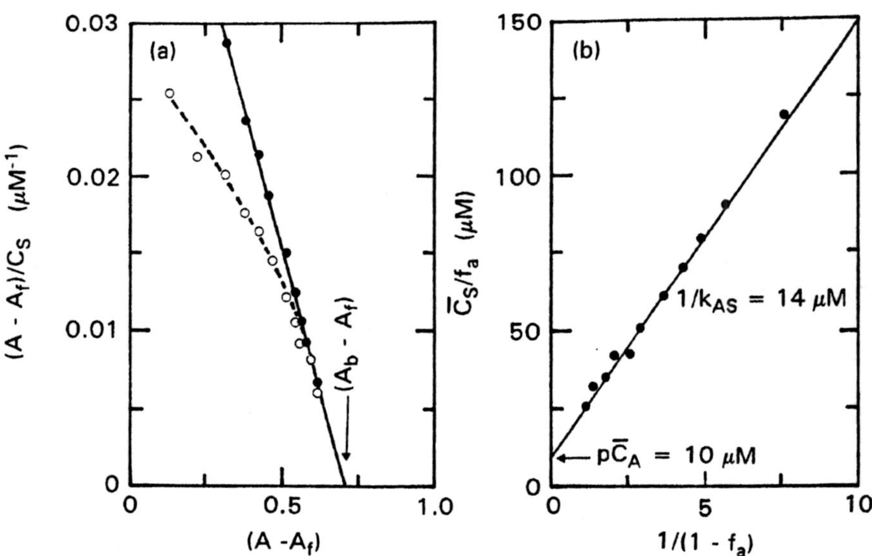

Fig. 3.5. Simultaneous evaluation of the stoichiometry and binding constant for the interaction of chloroquine (S) with uroporphyrin (A). **(a)** Scatchard plot of difference spectroscopy results with 10 μM acceptor, the open symbols denoting the plot obtained with total chloroquine concentrations $(\overline{C}_S)$ taken as initial estimates of C_S. **(b)** Secondary plot of the data in accordance with eq. 3.7b to obtain the intrinsic binding constant and stoichiometry from the slope $(1/k_{AS})$ and ordinate intercept $(p\overline{C}_A)$, respectively. [Adapted from Shanley *et al.* (1985).]

or

$$\overline{C}_S/f_a = p\overline{C}_A + 1/[k_{AS}(1 - f_a)] \tag{3.7b}$$

A secondary plot of the results in terms of eq. 3.7b may therefore be used (Fig. 3.5b) to obtain an initial estimate of the stoichiometry. A value of unity is indicated by the ordinate intercept of 10 μM in an experiment with $\overline{C}_A$ = 10 μM. Such evaluation of p now allows refinement of the values of C_S (eq. 3.3) and hence iterative refinement of the interaction parameters p and k_{AS}. In this particular instance (Fig. 3.5) the use of $\overline{C}_S - f_a\overline{C}_A$ as the second estimate of C_S led to no change in the maximal absorbance change $[A_b - A_f = (y_b - y_f)\overline{C}_A]$ and hence no change in f_a. It may therefore be concluded that the interaction of chloroquine with uroporphyrin is characterized by 1 : 1 stoichiometry and a binding constant of 7,000 M^{-1} (Shanley *et al.*, 1985).

A special case worthy of comment concerns the characterization of relatively weak binding by titrations in which the spectral parameter is a property of the acceptor. In these situations the inequality $\overline{C}_S >> p\overline{C}_A$ is likely to pertain, and hence the free concentration of ligand (C_S) may reasonably be approximated by $\overline{C}_S$. Such conditions improve the accuracy with which y_b may be determined, because the initial Scatchard plot of the results with $\overline{C}_S$ as the estimate of C_S is linear. Indeed, the slope of the plot also defines k_{AS}. However, no information on the magnitude of p emanates from the secondary plot in these circumstances, because the ordinate intercept ($p\overline{C}_A$) becomes indistinguishable from zero—a factor evident from eq. 3.7 on noting that $\overline{C}_S - pf_a\overline{C}_A \approx \overline{C}_S$. Such a situation is frequently encountered in studies of xenobiotic interactions with the microsomal cytochrome P-450 system (Fig. 3.1).

3.5. ISOPARAMETRIC ANALYSIS

In the formulation of eqs. 3.1 and 3.5 it was assumed that the binding of each ligand molecule to acceptor results in an equivalent change in the spectroscopic signal. This need not be the case, especially if the ligand binding is to multiple, nonequivalent sites. The assumption can be verified by showing that under conditions of stoichiometric titration the spectral change varies linearly with the binding function (or fractional saturation). In some cases the nonlinearity of this plot may be a consequence of the spectroscopic method used rather than nonequivalence of the spectroscopic signal for each site occupied. For example, the binding of NADH to lactate dehydrogenase results in quenching of tryptophan fluorescence due to resonance energy transfer from tryptophan to NADH. However, the quenching increment is not the same for each site occupied because the binding of NADH to one subunit of the tetrameric enzyme not only quenches a tryptophan residue within that subunit but also quenches the fluorescence of a tryptophan in an adjacent subunit. Consequently, when the coenzyme-binding site in the latter subunit is eventually occupied, its quenching increment appears to be smaller even though the binding of NADH to the four enzyme sites (one in each subunit) is equivalent and independent (Holbrook,

1972). A similar effect, which reflects distance dependence of the Forster energy transfer process, is also encountered in the binding of fluorescent fatty acids (*e.g.*, *cis*-parinaric acid) to human serum albumin and α-fetoprotein (Sklar *et al.*, 1977; Berde *et al.*, 1979). For these two systems the enhancement of ligand fluorescence was used to characterize the interactions (binding constant and stoichiometry). However, the binding also resulted in quenching of the single tryptophan residue in human serum albumin due to Forster-type energy transfer. Furthermore, the degree of tryptophan quenching did not follow the degree of fluorescence enhancement, since each site occupied was a different distance from the tryptophan residue. Fluorescence resonance energy transfer is therefore of limited use as a spectral parameter for the quantitative measurement of binding. It is, however, an extremely useful "spectroscopic ruler" for the measurement of distances in macromolecules or large molecular assemblies.

In this section an approach is developed that avoids the restrictive assumption that the binding of each ligand elicits an equivalent change in the spectroscopic parameter being measured. It may be applied to systems in which the spectroscopic signal arises from either the acceptor or the ligand, and may also be applied to the analysis of partition equilibria and for distinguishing between binding and partitioning processes. The method has been termed *isoparametric analysis* because it involves analysis at a single level of an experimentally measured parameter (Chatelier and Sawyer, 1987).

In this procedure a ligand-addition titration is repeated for several fixed concentrations ($\overline{C}_A$) of acceptor. For each acceptor concentration the spectroscopic parameter (or function thereof) is plotted against $\overline{C}_S$, the total concentration of ligand (Fig. 3.6). If the signal arises from the acceptor, the ordinate may be expressed as fractional saturation. At a given concentration of total ligand the fractional saturation decreases with increasing acceptor concentration, despite the increasing amount of ligand bound. At a single value, f_a, of the fractional saturation (horizontal line, Fig. 3.6a) there is a unique value of the binding function (r) across all values of $\overline{C}_A$, these two parameters being related by the expression

$$(\overline{C}_S)_{f_a} = C_S + r\overline{C}_A \tag{3.8}$$

Thus a secondary plot of $(\overline{C}_S)_{f_a}$ *versus* $\overline{C}_A$ (Fig. 3.6b) is linear with slope r and ordinate intercept C_S. Repetition of this analysis at several levels of fractional saturation permits the construction of a binding curve and/or a linear transform thereof.

The first point to note in relation to the isoparametric analysis is that the primary plot in Figure 3.6 is not a true binding curve because the abscissa is expressed in terms of $\overline{C}_S$ instead of C_S, this factor being the cause of the dependence of f_a on acceptor concentration. Second, the ordinate need not necessarily be expressed in terms of the fractional saturation, a normalized curve of the observed spectroscopic parameter being sufficient. For example, in an experiment with signal originating from the acceptor, the ordinate can be

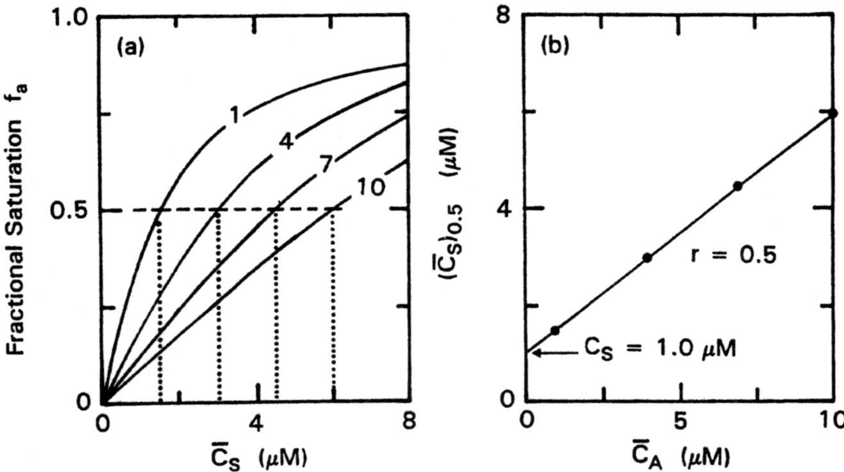

Fig. 3.6. Illustration of the isoparametric analysis. **(a)** Simulated titration curves for an acceptor–ligand system with $p = 1$, $k_{AS} = 10^6\ \mathrm{M}^{-1}$, and the indicated total concentrations (μM) of acceptor ($\overline{C}_A$). Vertical lines denote the values of $\overline{C}_S$ for which the fractional saturation (f_a) is 0.5. **(b)** Secondary plot of these $(\overline{C}_S)_{0.5}$ values in accordance with eq. 3.8 to obtain the binding function (r) and the associated free ligand concentration (C_S) from the slope and ordinate intercept, respectively.

scaled as $\Delta y/\overline{C}_A$, where Δy is the observed change in spectroscopic signal ($y - y_f$). Finally, the method does not, in the first instance, require the assumption of a binding model. Rather, its aim is to provide an alternative means of defining the dependence of the binding function upon free ligand concentration—a relationship that may then be tested for conformity with any proposed model of the interaction.

An unusual application of isoparametric analysis is the study of the binding of fluoresceinated epidermal growth factor to cell surface receptors on a cultured cell line by flow cytometry (Chatelier *et al.*, 1986). In this technique the fluorescence signal arises solely from the bound dye, since there is rapid (10 ms) dilution of the unbound dye by the sheath fluid as it mixes with the cell stream. The titration was carried out at several cell densities (Fig. 3.7a) and the data analyzed in accordance with eq. 3.8 (Fig. 3.7b) to obtain the binding curve (r as a function of C_S), which indicated the presence of two classes of independent cell surface receptor sites for epidermal growth factor (Chatelier *et al.*, 1986). The method has been applied to the binding of a protein ligand to DNA (Lohman and Mascotti, 1992).

3.6. NMR AND ESR STUDIES OF LIGAND BINDING

Nuclear magnetic resonance (NMR) and electron spin resonance (ESR) are spectral techniques that may also be employed to characterize acceptor–ligand

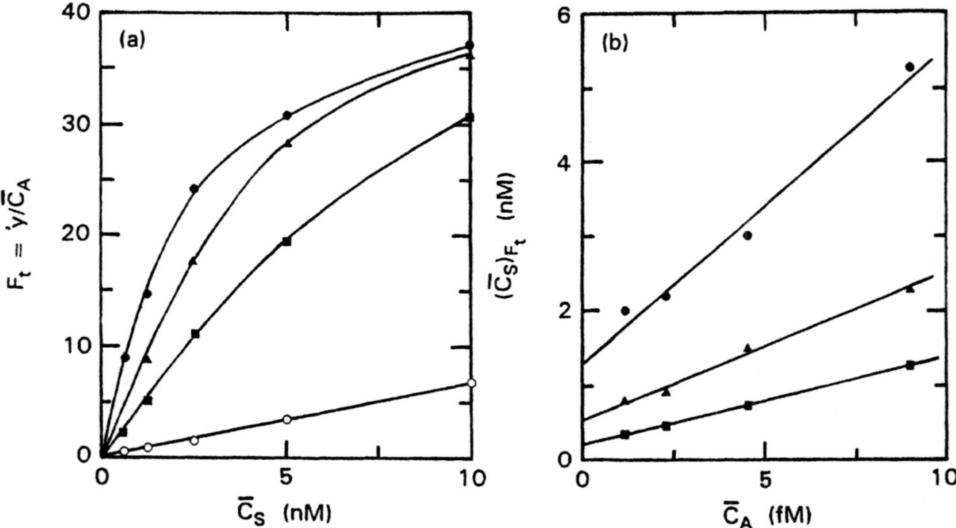

Fig. 3.7. Isoparametric analysis of the binding of fluorescein-labeled epidermal growth factor (S) to A431 cells (A). **(a)** Titration curves, fluorescence per cell (F_t) *versus* C_S, obtained by quantitative flow cytometry in experiments with 7 × 10^5 cells/ml (•), 2.7 × 10^6 cells/ml (▲), and 5.4 × 10^6 cells/ml (■). The results have been corrected for autofluorescence of the cells but require correction for nonspecific binding. Open symbols denote the fluorescence due to nonspecific binding in the titration with lowest cell concentration. **(b)** Secondary plots of data in accordance with eq. 3.8 for fixed F_t values of 20 (•), 15 (▲), and 10 (■). [Adapted from Chatelier *et al.* (1986).]

interactions, the major proviso being that equilibrium attainment is again sufficiently rapid for the spectrum to contain a composite peak that reflects the equilibrium between free and bound states of one component of the mixture (Dwek, 1973; Berliner, 1981). The most common method of characterizing acceptor–ligand interactions entails analysis of the composition dependence of a chemical shift in the spectrum of either the acceptor (Meadows and Jardetzky, 1968; Griffin *et al.*, 1973) or the ligand (Spotswood *et al.*, 1967; Raftery *et al.*, 1968). Alternatively, the corresponding dependence of the spread of specific ligand peaks (Berliner and Wong, 1975) or of rates (spin-lattice or spin-spin) of nuclear dipole relaxation (James and Noggle, 1969; Lanir and Navon, 1971; Nowak, 1981) may be used to characterize an interaction. An example of complex formation being manifested as a chemical shift in the ligand spectrum is shown in Figure 3.8a for interaction of the D-isomer (but not the L-isomer) of *N*-acetyl-*p*-fluorophenylalanine with α-chymotrypsin (Spotswood *et al.*, 1967). On the other hand, Figures 3.8b,c provide examples where complex formation gives rise to peak-broadening in NMR and ESR spectra, respectively.

The observed average spectral parameter (chemical shift, peak height, or relaxation rate) is, of course, amenable to analysis by the procedures described

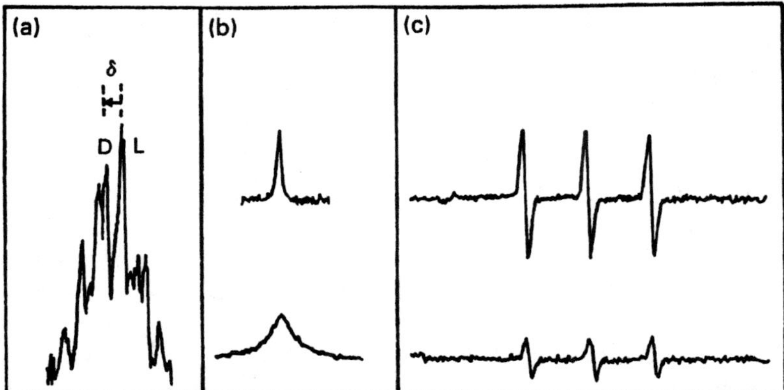

Fig. 3.8. Different methods of detecting ligand interactions by NMR and ESR spectroscopy. **(a)** Chemical shift (δ) in the NMR spectrum of *N*-acetyl-DL-*p*-fluorophenylalanine (83 mM) in the presence of α-chymotrypsin (3 mM): the D-isomer is the reactive species. [Adapted from Spotswood *et al.* (1967).] **(b)** Peak broadening in the NMR spectrum of *N*-acetylsulfonamide (20 mM) in the presence of carbonic anhydrase (1.47 mM). [Adapted from Lanir and Navon (1971).] **(c)** Decreased peak height resulting from broadening of the ESR spectrum of a nitroxide derivative of UDP (13.9 μM) in the presence of galactosyl transferase (1.09 mM). [Adapted from Berliner and Wong (1975).]

in Sections 3.2 and 3.3. However, the relative insensitivity of NMR in comparison with the detection capabilities of its spectrophotometric and fluorimetric counterparts often leads to a situation where the binding constant is being determined from data reflecting relatively high saturation of acceptor. Prior titration of acceptor under stoichiometric conditions to determine the spectral characteristic of fully saturated acceptor–ligand complex, AS_p, is then unnecessary.

3.6.1. Binding Constant for a Weak Interaction

Insensitivity of the NMR spectral signal dictates the use of high ligand concentrations and hence restricts its application to the quantitative characterization of relatively weak interactions. Because of the consequent likelihood that $C_S \approx \overline{C}_S$, we choose to write the rectangular hyperbolic expression for ligand binding in the form

$$r = \frac{f_s\overline{C}_S}{\overline{C}_A} = \frac{(y - y_f)\overline{C}_S}{(y_b - y_f)\overline{C}_A} = \frac{pk_{AS}\overline{C}_S(C_S/\overline{C}_S)}{1 + k_{AS}\overline{C}_S(C_S/\overline{C}_S)} \tag{3.9}$$

where a value of unity may reasonably be assigned to the ratio $C_S/\overline{C}_S$ if $\overline{C}_S >> p\overline{C}_A$ and where either the chemical shift (δ) or relaxation rate ($1/T_1$ or $1/T_2$) may be substituted for y in the expression for f_s, the fraction of ligand bound (eq. 3.4). For experiments performed with constant acceptor concentration ($\overline{C}_A$) the results have sometimes been analyzed in terms of the double-reciprocal linear transform of eq. 3.9 with $C_S/\overline{C}_S = 1$ (*e.g.*, Spotswood *et al.*, 1967; Sykes, 1969). However, we note that a more convenient linear transform of eq. 3.9, namely,

$$(y - y_f)/(C_S/\overline{C}_S) = pk_{AS}\overline{C}_A(y_b - y_f) - k_{AS}\overline{C}_S(y - y_f) \qquad (3.10)$$

allows direct evaluation of k_{AS} from the slope of a plot of $(y - y_f)$ *versus* $\overline{C}_S(y - y_f)$ in the event that $C_S \approx \overline{C}_S$. Furthermore, should retrospective considerations reveal this assumption to be untenable, the resulting value of $y_b - y_f$ that is obtained from the ordinate intercept for an interaction with defined stoichiometry (p) provides a value of y_b and hence a means of refining the estimate of $C_S/\overline{C}_S = (y_b - y)/(y_b - y_f)$ to be used in eq. 3.9 or 3.10.

Iterative application of eq. 3.10 to chemical shift data for results of an NMR investigation of the interaction between α-chymoptrypsin and trifluoroacetyl-D-phenylalanine (Sykes, 1969) is illustrated in Figure 3.9a. The 2.5–4.0% differences between C_S and $\overline{C}_S$ inferred from the initial analysis do lead to slightly revised estimates of k_{AS} (20 ± 5 *cf.* 19 ± 5 M^{-1}) and $\delta_b - \delta_f$ (41 ± 7 *cf.* 42 ± 7), but no statistical significance can be attached to the differences. Although it may therefore be argued that the iterative analysis serves no useful purpose, such unequivocal justification of the approximation that $C_S \approx \overline{C}_S$ could well be important in studies of stronger interactions by NMR or ESR (James and Noggle, 1969).

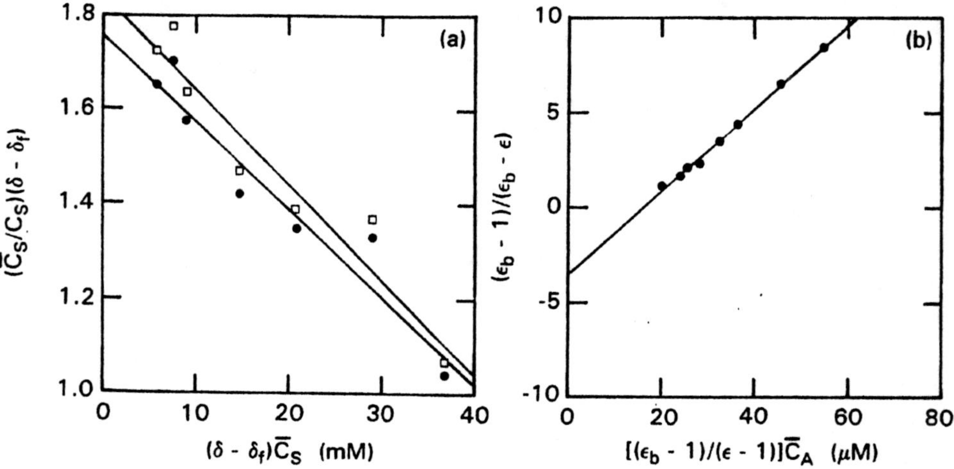

Fig. 3.9. Determination of binding constants by NMR and ESR spectroscopy. **(a)** Iterative application of eq. 3.10 to NMR chemical shift data for the interaction of trifluoroacetyl-D-phenylalanine with α-chymotrypsin (Sykes, 1969): •, □, refer to the initial and revised estimates, respectively, of the ordinate parameter. **(b)** Application of eq. 3.11 to the enhancement of spin-lattice relaxation rate ($1/T_1$) associated with complex formation between Mn^{2+} and enolase (Nowak, 1981).

3.6.2. Analysis of Mixtures With Fixed Ligand Concentration

The experimental design for many characterizations of ligand binding departs from our recommended procedure in that the results are obtained by varying the total acceptor concentration in mixtures containing a fixed total concentration of ligand that is optimal for monitoring by the particular spectroscopic technique—be it spectrophotometry, fluorimetry, NMR, or ESR. As detailed elsewhere (Chapter 6), this procedure only yields a meaningful binding curve if self-interaction of the acceptor is either nonexistent or restricted to isomerization. Although results obtained by this means are usually interpreted by calculating f_s (eq. 3.4) to obtain C_S and r for conventional Scatchard analysis, such action is misleading in the sense that it provides no indication of the fact that $\overline{C}_A$ (not $\overline{C}_S$) is the quantity being varied in the mixtures under examination. We therefore introduce a linear transform of the rectangular hyperbolic expression for ligand binding that highlights this distinction.

Equation 3.9 continues to apply, but account now needs to be taken of the fact that neither $\overline{C}_A$ nor $C_S/\overline{C}_S$ can be approximated by constants. Provided that y_b is obtainable by extrapolation of data to infinite acceptor concentration, $C_S/\overline{C}_S = 1 - f_s$ can be replaced by $(y_b - y)/(y_b - y_f)$, which allows rearrangement of eq. 3.9 as

$$(y_b - y_f)/(y_b - y) = -k_{AS}\overline{C}_S + pk_{AS}\overline{C}_A(y_b - y_f)/(y - y_f) \quad (3.11)$$

Equivalent and independent binding of ligand to acceptor is thus recognized by a linear dependence of $(y_b - y_f)/(y_b - y)$ *versus* $\overline{C}_A(y_b - y_f)/(y - y_f)$, whereupon the intrinsic binding constant may be deduced from either the ordinate intercept or the slope.

The results chosen for illustrative purposes emanate from an NMR characterization of the interaction between Mn^{2+} and enolase on the basis of the enhancement of the spin–lattice relaxation rate ($1/T_1$) associated with complex formation (Nowak, 1981). Because the relaxation rate enhancement was expressed relative to that for free ligand ($\epsilon = y/y_f$), the plot of the results (Nowak, 1981) in accordance with eq. 3.11 incorporates this change (Fig. 3.9b). Interpretation of the ordinate intercept as $-k_{AS}\overline{C}_S$ for these experiments with 40 μM Mn^{2+} yields a binding constant of 90,000 ($\pm$ 20,000) M^{-1} for the Mn^{2+}–enolase interaction, essentially the same value, 110,000 ($\pm$10,000) M^{-1} being obtained on the basis of the slope and a value of 2 for the valence of this dimeric enzyme.

Retrospective inspection of the procedures outlined in this chapter reveals common links between the various methods of characterizing acceptor–ligand interactions by spectroscopic techniques. Irrespective of the technique being used, the analysis is based on the experimental determination of either f_a, the fractional saturation of acceptor sites, or f_s, the fraction of ligand bound to acceptor. In all procedures this information has been derived from a specific spectral feature that reflects the distribution of one component (either acceptor or ligand) between free and complexed states. As stated explicitly in this final section on NMR and ESR, analyses of spectral data are invariably based on the premise that equilibrium attainment is essentially instantaneous on the time-scale of the experimental measurements. Only in studies involving measurement of nuclear dipole relaxation rates is the validity of this presumption likely to come into question.

4

OTHER DIRECT PROCEDURES FOR CHARACTERIZING LIGAND BINDING

This chapter describes a miscellany of procedures that may be used to characterize ligand binding under specific circumstances. Of these, the first relies upon the availability of a method for direct measurement of the concentration of free ligand in the equilibrium mixture, whereas the second and third depend upon the ability to observe the changed macromolecular status of a ligand associated with acceptor, or of an acceptor associated with ligand. Finally, mention is made of microcalorimetry, which is gaining popularity as a method for quantifying interactions that are characterized by a finite enthalpy change ($\Delta H \neq 0$).

4.1. SPECIFIC ION ELECTRODES

Any requirement for phase separation (as in Chapter 2) is clearly rendered redundant by the availability of an electrode that responds solely to the concentration of free ligand in an acceptor–ligand mixture. The best-known specific ion electrode is, of course, the glass electrode, which has been used widely to quantify proton binding to proteins and other macromolecular ions. Other specific electrodes for particular ions are also available, but a major limitation of their use for characterizing ligand binding is their sensitivity, which is seldom better than 1 μM. This sensitivity restriction is no impediment to quantification of relatively weak interactions, as is evident from studies of the binding of Ca^{2+} to monophosphates and *myo*-inositol polyphosphates (Luttrell, 1993), for which the dissociation constants are in the 10 mM to 10 μM range. Here, we focus on the determination and interpretation of pH-titration curves to exemplify the type of information that can emanate from measurement of free ligand concentration by means of a specific ion electrode.

4.1.1. Hydrogen Ion Titrations: The Proton as Ligand

Binding equilibria of particular interest to biochemists involve the dissociation of a proton from a weak acid and the association of a proton with a weak base. Not only do these equilibria form the basis of buffers used almost universally in biochemical experiments, but they also have a profound effect on the physical properties of proteins due to the presence of ionizable side chains on the amino acid residues comprising the polypeptide chain. In this section we treat the hydrogen ion as a ligand that binds to multiple nonequivalent sites on a protein acceptor.

The concentration of free ligand is most readily measured with a pH electrode, and hence the titration curve of the protein becomes a binding curve, with the charge on the protein reflecting the number of protons bound and the pH indicating the concentration of free protons in solution. The titration curve can therefore be approximated as the sum of the binding curves that describe the binding equilibria of the individual titratable groups, for which the binding constants depend upon the chemical nature of the ionizable group and the characteristics of its local environment. An unusual feature of these titration curves is the enormous range of binding affinities that is covered—a consideration that dictates that the concentration of free protons be varied between 10^{-2} and 10^{-14} M and hence that the abscissa of the binding curve be expressed on a logarithmic scale (pH). For the same reason the dissociation constants are frequently described in terms of their negative logarithms, *i.e.*, $pK = -\log k_d$. By analogy with eq. 1.11, the binding of hydrogen ions to i sets of equivalent and independent protein sites, each set containing n_i titratable groups, is described by the expression

$$r = \Sigma \, \{n_i(k_{AS})_i C_{\mathrm{H}^+}/[1 + (k_{AS})_i C_{\mathrm{H}^+}]\} \qquad (4.1)$$

or, because ionization constants are usually expressed as dissociation constants, k_d,

$$r = \Sigma \, \{n_i C_{\mathrm{H}^+}/[(k_d)_i + C_{\mathrm{H}^+}]\} \qquad (4.2)$$

Not all ionizable groups of a compact, globular protein are accessible to the bulk solvent and hence available for titration. Only in the presence of a strong denaturant such as 6 M guanidine hydrochloride is it reasonable to assume that all ionizable groups are available for titration. Consequently, a comparison of titration curves of the native and denatured protein provides a measure of the number of ionizable groups that are buried in the interior of the native protein.

The first step in the experimental determination of a pH-titration curve is to prepare a sample of protein that is free of any bound anions or cations—a task that is accomplished by passing the protein solution through a mixed-bed ion-exchange column. The resulting solution is then supplemented with electrolyte (usually KCl) or with electrolyte and denaturant, after which either acid or alkali is added. For illustration purposes, we shall consider that acid is being added. The measured activity of hydrogen ion (10^{-pH}) after each addition is divided by the activity coefficient (γ_{HCl}) and then subtracted from the total added hydrogen ion concentration to obtain the concentration of bound protons. Division by the total protein concentration ($\overline{C}_A$) then gives the binding function (r).

Frequently, the ordinate of a pH-titration curve is expressed in terms of net protein charge (valence), which is obtained by subtracting from r the number of protons bound to the protein at its isoionic pH. An obvious complication inherent in the method is that the binding function is defined in terms of concentrations, whereas the glass electrode provides a measure of the thermodynamic activity of hydrogen ion. The experiment is therefore conducted at fixed ionic strength (by inclusion of, *e.g.*, 0.2 M KCl) so that the required activity coefficient (γ_{HCl}) can be obtained from tables supplied by Bates (1964). Detailed practical and theoretical descriptions of the procedure are given by Tanford (1961) and Steinhardt and Reynolds (1969).

The titration curve of ribonuclease in 6 M guanidine hydrochloride (Nozaki and Tanford, 1967), shown in Figure 4.1a, establishes that this enzyme can bind a total of 32 protons in the pH range 2–12, which corresponds to the number of titratable groups present (5 asp, 5 glu, 4 his, 10 lys, 6 tyr, and the α-amino and the α-carboxyl groups). The 4 arg residues do not titrate in this pH range. The solid line in Figure 4.1a represents the titration curve generated by incorporating the appropriate values of n_i and $(k_d)_i$ into eq. 4.2. In that regard, the tabulated dissociation constants (Table 1 of Nozaki and Tanford, 1967) refer to proton ionization under conditions of zero net charge. The required value is related to that standardized parameter, k_d^o, by the expression

$$(k_d)_i = (k_d^o)_i \exp(2wZ) \tag{4.3a}$$

Z is the net charge, and w, the electrostatic interaction factor (Linderstrom-Lang, 1924; Tanford, 1950), is given by

$$w = (e^2/DR_AkT)[1 + \kappa R_A/(1 + \kappa R_A + \kappa R_i)] \tag{4.3b}$$

where e is the electronic charge, D is the dielectric constant, and κ, the inverse screening length, may be calculated from the ionic strength (I) as $3.27 \times 10^9 \sqrt{I}$ (the value for uni-univalent electrolytes at 25°C). R_A and R_i are the respective radii of the protein and the average supporting electrolyte ion (approx. 0.2 nm). From the experimental viewpoint the electrostatic interaction factor is initially evaluated as a curve-fitting parameter whose magnitude is then checked for conformity with eq. 4.3b. However, in 6 M guanidine hydrochloride (the conditions prevailing in Fig. 4.1a), the extremely high ionic strength decreases the electrostatic factor to effectively zero, and hence $(k_d)_i = (k_d)_o$ (Nozaki and Tanford, 1967).

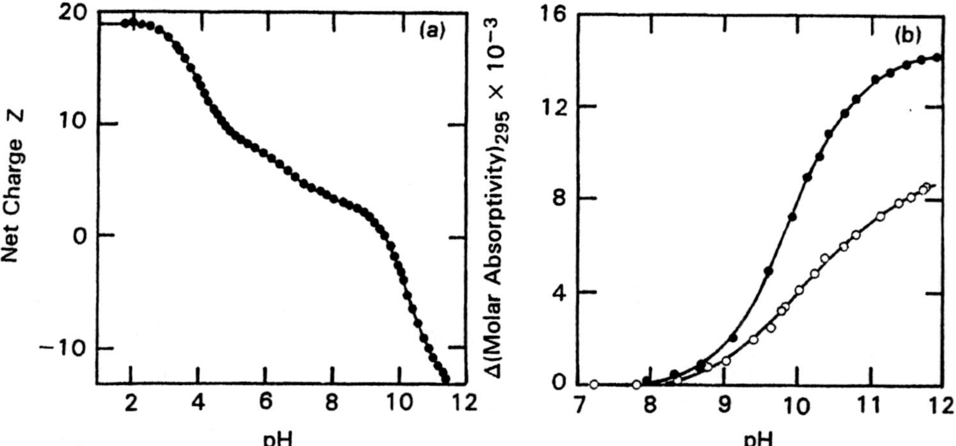

Fig. 4.1. Proton binding by ribonuclease. **(a)** pH titration curve of the enzyme in 6 M guanidine hydrochloride, together with its theoretical description (line) in terms of group numbers and the pk_d values reported by Nozaki and Tanford (1967). **(b)** Comparison of the spectrophotometric titrations of the tyrosine residues of ribonuclease in the presence of 6 M guanidine hydrochloride (●) with that determined for native enzyme (○). [Adapted from Nozaki and Tanford (1967) and Tanford *et al.* (1955), respectively.]

Spectrophotometric titration of the protein provides an independent monitor of tyrosyl group ionization, which is accompanied by an increase in the absorption coefficient of the phenolic hydroxyl (Tanford *et al.*, 1955). This procedure has been used to demonstrate that the three additional proton ionizations observed in the titration of denatured ribonuclease reflect tyrosine residues that do not ionize in the native enzyme (Fig. 4.1b).

It is clearly an approximation to assume that each titratable group of a particular class (*e.g.*, carboxyl groups) exhibits the same pk_d value, because each is in a unique environment within the native structure and therefore likely to possess unique ionization characteristics. However, the hydrogen-ion titration curve is rarely able to resolve these individual pk_d values because of the overlapping titration of each group. For histidine residues, proton NMR can sometimes be used to resolve and identify each titratable group, in which case the individual titration curves can be followed independently. Such titration of the four histidine residues of ribonuclease has indeed been reported (Meadows *et al.*, 1968).

4.2. MEASUREMENT OF CONSTITUENT RATES OF MIGRATION

In mass migration experiments such as electrophoresis, sedimentation velocity, and chromatography, the interaction of ligand with acceptor can frequently be detected by a change in the migration rate of either the ligand or the acceptor constituent. For example, in electrophoresis the binding of a small metal ion to a negatively charged protein yields a protein–ligand complex that is more electropositive than the free protein and therefore is characterized by a decreased cathodic mobility.

Provided that chemical equilibrium attainment is rapid on the time-scale of acceptor and ligand separation by the particular experimental method being used, the migration can be described in terms of average velocities of the acceptor and ligand components. These two parameters, termed *constituent velocities* ($\overline{v}_i$; $i = A$ or S), are given by

$$\overline{v}_A = \left[v_A C_A + \sum_1^p (v_{AS_i} C_{AS_i}) \right] \Big/ \overline{C}_A \tag{4.4a}$$

$$\overline{v}_S = \left[v_S C_S + \sum_1^p (i v_{AS_i} C_{AS_i}) \right] \Big/ \overline{C}_S \tag{4.4b}$$

for a system in which accpetor A, characterized by velocity v_A and present at total molar concentration $\overline{C}_A$, possesses p sites for interaction with ligand S, characterized by velocity v_S and present at total molar concentration $\overline{C}_S$. For mixtures with a fixed total concentration of acceptor, $\overline{C}_A$, and a range of total ligand concentrations, $\overline{C}_S$, the constituent velocity of acceptor varies progressively from v_A to v_{AS_p} as $\overline{C}_S$ is increased from zero to a concentration that suffices to saturate all p sites on A. Measurement of $\overline{v}_A$ as a function of $\overline{C}_S$ thus has the potential to provide quantitative information on the acceptor–ligand interaction. On the other hand, if the constituent velocity of ligand were more amenable to experimental measurement than $\overline{v}_A$, a corresponding variation of $\overline{v}_S$ with $\overline{C}_A$ would be observed in experiments on mixtures with fixed $\overline{C}_S$ and a range of $\overline{C}_A$ values.

4.2.1. Characterization of Single Complex Formation

If the acceptor possesses only a single site for ligand ($p = 1$), eq. 4.4a becomes

$$\overline{v}_A = [v_A C_A + v_{AS}(\overline{C}_A - C_A)]/\overline{C}_A \tag{4.5}$$

on the grounds that C_{AS} is simply the difference between the total and free concentrations of acceptor. By simple algebraic manipulation of eq. 4.5, it then follows that

$$C_A = \overline{C}_A(\overline{v}_A - v_{AS})/(v_A - v_{AS}) \tag{4.6a}$$

$$C_{AS} = (\overline{C}_A - C_A) = \overline{C}_A(v_A - \overline{v}_A)/(v_A - v_{AS}) \tag{4.6b}$$

The binding constant may thus be determined from the consequent values of C_{AS} and C_A as a function of $\overline{C}_S$, because

$$(\overline{C}_A - C_A)/C_A = k_{AS}C_S = k_{AS}(\overline{C}_S - \overline{C}_A + C_A) \tag{4.7}$$

Use of this approach to evaluate a binding constant is illustrated in Figure 4.2, which summarizes a gel chromatographic study of complex formation between thiamin diphosphate and Mg^{2+} (Booth *et al.*, 1992). From Figure 4.2a, which compares advancing elution profiles for the thiamin diphosphate constituent (A) in the absence of metal ion (——) and in the presence of a saturating Mg^{2+} concentration (– · – · –), it is evident that the complex exhibits a larger elution volume than the nucleotide. For a lower concentration of metal ion, the thiamin diphosphate migrates with an intermediate constituent velocity (– – –). In the application of eqs. 4.6a and 4.6b to obtain C_A and C_{AS} from such data, it is noted that the chromatographic analog of velocity in an elution profile is the elution volume (Ackers and Thompson, 1965; Gilbert, 1966; Nichol *et al.*, 1967c). From Figure 4.2b, which analyzes, in accordance with eq. 4.7, the results from experiments with a range of metal ion concentrations, the interaction of Mg^{2+} with thiamin diphosphate is characterized by a binding constant of 3,000 M^{-1} under the conditions investigated (0.1 M Tris/HCl, pH 7.6).

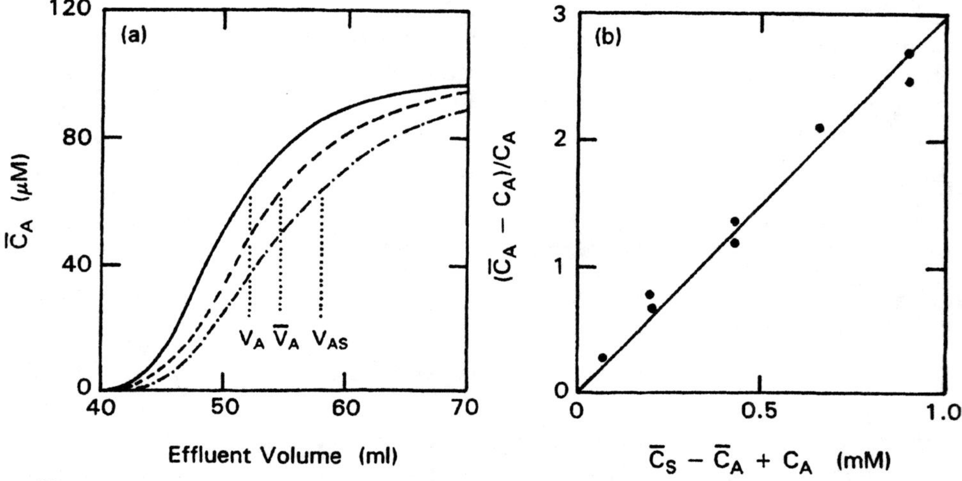

Fig. 4.2. Characterization of the interaction between thiamin diphosphate and Mg^{2+} by gel chromatography. **(a)** Advancing elution profiles for thiamin diphosphate (100 μM) on a column of Sephadex G-10 in the absence of metal ion (solid line) and in the presence of 0.245 mM (dashed line) and 100 mM $MgCl_2$ (dashed, dotted line). [Adapted from Booth *et al.* (1992).] **(b)** Plot of the results [from Table 1 of Booth *et al.* (1992)] in accordance with eq. 4.7 to obtain k_{AS} as the slope.

4.3.2. Characterization of Multiple Complex Formation

If $p > 1$, an analytical solution to eq. 4.4 requires specification of v_{AS_i}, the velocity of each acceptor–ligand species, AS_i ($1 \leq i \leq p$). Consequently, even under circumstances where all interactions between acceptor and ligand are governed by a single intrinsic binding constant, the problem of determining the magnitude of k_{AS} is essentially intractable without invoking some formal relationship between the various v_{AS_i}. Provided that each successive binding of a charged ligand to a protein acceptor gives rise to a constant incremental change in electrophoretic mobility of acceptor (Fig. 4.3a), the constituent mobility ($\overline{v}_A$) is related to the mobility of free acceptor, v_A, in electrophoresis by the expression

$$(\overline{v}_A - v_A)/v_A = (p\delta)k_{AS}C_S/(1 + k_{AS}C_S) \tag{4.8}$$

in which $\delta = (v_{AS_i} - v_{AS_{i-1}})/v_A$ is the incremental change in mobility expressed as a fraction of the mobility of free A (Drewe and Winzor, 1976). The two parameters that emanate from the rectangular hyperbolic dependence of $(\overline{v}_A - v_A)/v_A$ upon C_S thus define the binding constant, k_{AS}, and $p\delta$, the mobility difference between v_A and v_{AS_p} as a fraction of v_A. This approach to the evaluation of a binding constant from electrophoretic data reflecting the interaction of phosphate with ovalbumin is illustrated in Figure 4.3b, which combines

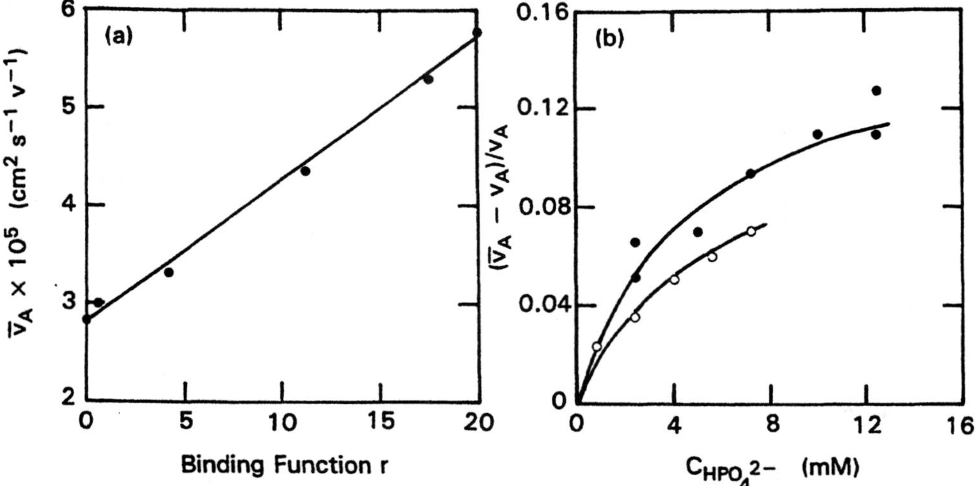

Fig. 4.3. Use of constituent rates of electrophoretic migration to characterize the interactions of a charged ligand with multiple acceptor sites. **(a)** Linear dependence of the electrophoretic mobility ($\overline{v}_A$) of bovine serum albumin upon the extent of site saturation with methyl orange. [Adapted from Smith and Briggs (1950).] **(b)** Application of eq. 4.8 to results obtained by moving boundary electrophoresis (○) and polyacrylamide gel electrophoresis (●) of ovalbumin–phosphate mixtures. [Adapted from Drewe and Winzor (1976) and Ward and Winzor (1981), respectively.]

results obtained by moving boundary electrophoresis (Drewe and Winzor, 1976) and also by a quantitative adaptation of polyacrylamide gel electrophoresis (Ward and Winzor, 1981). A binding constant of 200 M^{-1} describes both sets of results, but the value of $p\delta$ is smaller in the latter study, performed at a higher ionic strength (**I** 0.15 *cf.* 0.10).

4.2.3. Affinity Electrophoresis

The technique of affinity electrophoresis (Takeo and Nakamura, 1972) also relies upon measurement of the constituent velocity of one reactant (acceptor) in the presence of a range of concentrations of another reactant (ligand). Polyacrylamide gel electrophoresis was used to study the interaction between muscle phosphorylase and glycogen, a ligand with zero velocity because of its inability to migrate through the gel pores. Consequently, the enzyme–glycogen complexes were also stationary ($v_{AS_i} = 0$), in which case eq. 4.5 simplifies to

$$\overline{v}_A = v_A C_A / \overline{C}_A \tag{4.9a}$$

or

$$C_A = \overline{C}_A (\overline{v}_A / v_A) \tag{4.9b}$$

For details of the application of eq. 4.9 to those results for the phosphorylase–glycogen system, the reader is referred to a previous review article (Winzor and de Jersey, 1989).

An important point to note is that a complex with zero velocity is also formed between the partitioning solute and the immobilized reactant in affinity chromatography. Indeed, eq. 4.9 plays an important role in the characterization of interactions by quantitative affinity chromatography—a topic to be considered in the context of characterizing acceptor–ligand equilibria by competitive binding assays (Chapter 5).

4.3. ANALYTICAL ULTRACENTRIFUGATION

In principle, ultracentrifugation holds considerable promise as a means of characterizing interactions between dissimilar reactants, particularly in instances where acceptor and ligand are both macromolecular. However, this potential is yet to be realized because research emphasis thus far has been on the evaluation of binding constants for small ligands—an application that does not allow full advantage to be taken of the power of sedimentation techniques for the quantitative characterization of interactions involving macromolecules.

4.3.1. Sedimentation Velocity

In this form of analytical ultracentrifugation, an equilibrium mixture of acceptor and ligand is subjected to a sufficiently high centrifugal field to generate a boundary between a plateau region with original composition and a ligand plateau region devoid of acceptor constituent. Although the rate of migration of this boundary may also be used to define the constituent velocity of the acceptor ($\overline{v}_A$), quantitative analysis based on eq. 4.5 is equivocal because of uncertainty about the magnitude to be assigned to the velocity of each complex species, v_{AS_i}. For small ligands it is usually a reasonable approximation to assume that ligand attachment is virtually without effect on the migration rate (sedimentation coefficient) of acceptor; but under those circumstances $\overline{v}_A = v_A$ and hence no binding information emanates from measurement of the constituent velocity. However, in such experiments the concentration of the ligand plateau behind the reaction boundary in acceptor constituent provides a direct measure of C_S in the original mixture with total ligand concentration $\overline{C}_S$ (see Section 2.4). This procedure has been used widely (*e.g.*, Chanutin *et al.*, 1942; Velick *et al.*, 1953; Bothwell *et al.*, 1978); but the validity of binding data so obtained is, of course, conditional upon the reliability of the inherent approximation that $v_{AS_i} = v_A$.

The advantage of the corresponding gel chromatographic techniques (Chapter 2) for obtaining the same information is that the required comigration of acceptor and all acceptor–ligand complexes can be effected readily and with certainty by selecting a gel matrix that excludes acceptor (and hence all acceptor–ligand complexes) from the stationary phase. No such guarantee pertains in sedimentation velocity measurements. Furthermore, as noted in Section 2.4.3 in relation to gel chromatography, the form of the sedimentation velocity pattern, *viz.*, a faster migrating reaction boundary followed by a pure ligand boundary, is likely to prevail even when $v_{AS_i} \neq v_A$, the problem being that the concentration of the pure ligand phase can no longer be equated with C_S. Although evaluation of a binding constant under those conditions is still possible (Nichol and Winzor, 1964; Nichol *et al.*, 1967b; Gilbert and Kellett, 1971; Cann and Winzor, 1987), the procedure is too complicated for inclusion in a treatise on ligand-binding techniques that are relatively simple to apply and interpret.

4.3.2. Sedimentation Equilibrium

Equilibrium centrifugation (Van Holde and Baldwin, 1958) has more to offer than its velocity counterpart because there is no ambiguity about the molecular weight of a postulated acceptor–ligand species: $M_{AS_i} = M_A + iM_S$. Operation of the ultracentrifuge at a much lower angular velocity in sedimentation equilibrium experiments means that there is no separation of components as in sedimentation velocity. Instead, centrifugation at angular velocity ω leads to a separate equilibrium distribution of each species governed by the expression

$$a_i(x) = a_i(x_F) \exp [M_i(1 - \overline{v}_i\rho)\omega^2(x^2 - x_F^2)/2\mathbf{RT}] \quad (4.10)$$

where the thermodynamic activity of species i at radial distance x, $a_i(x)$, is related to its activity at a selected radial position (x_F) within the liquid column by an exponential term containing the molecular weight of the species (M_i), its partial specific volume ($\bar{v}_i$), and the difference between the squared radial distances: **R** is the gas constant, **T** the absolute temperature, and ρ the solvent density (Wills and Winzor, 1992; Wills *et al.*, 1993). Because the time required for attainment of sedimentation equilibrium is relatively long (12–16 hours), it is presumed that species coexisting at any radial distance x are in chemical equilibrium. This situation is illustrated in Figure 4.4 for a system with 1 : 1 complex formation ($p = 1$) between acceptor ($M_A = 60{,}000$) and ligand ($M_S = 20{,}000$) governed by a binding constant (k_{AS}) of 10^4 M^{-1}. Other parameters used in the simulation (via eq. 4.10 on the basis of thermodynamic ideality) are given in the legend. An important development in the analysis of sedimentation equilibrium distributions for interacting systems has been the introduction of the omega function (Milthorpe *et al.*, 1975; Nichol *et al.*, 1976), which provides access to the free concentration of the smallest macromolecular species (S in the present context) throughout the distribution.

Consider initially the situation for which the separate distributions of the two components, *i.e.*, $\bar{C}_A(x)$ and $\bar{C}_S(x)$ as functions of radial distance, are available from optical scans of the equilibrium distribution at different wavelengths (Steinberg and Schachman, 1966), this being the situation illustrated

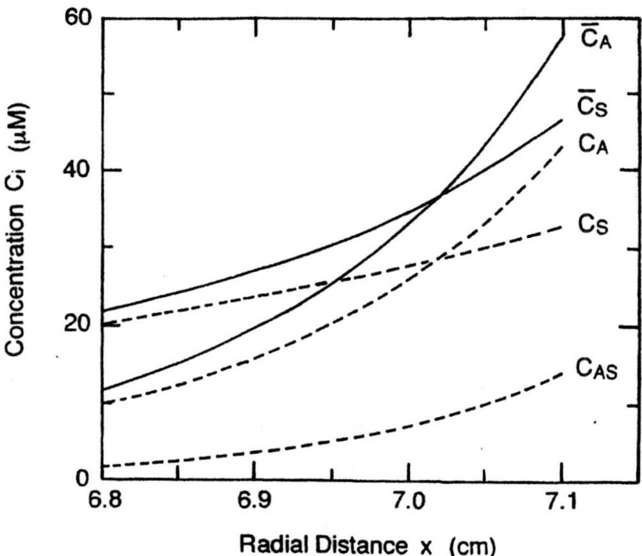

Fig. 4.4. Simulated sedimentation distributions for a mixture of acceptor ($M_A = 60{,}000$) and ligand ($M_S = 20{,}000$) undergoing reversible association to form a 1 : 1 complex. Distributions of individual species (dashed lines) are calculated from eq. 4.10 on the basis of respective values of 10 μM and 20 μM for C_A and C_S at the air–liquid extremity ($x = 6.80$ cm) of a 3-mm column subjected to centrifugation at 20°C and 10,000 rpm. The product $\bar{v}_{i\rho}$ has been taken as 0.260 for all species. Solid lines describe the radial dependence of the total (constituent) concentrations of A and S.

by the solid lines in Figure 4.4. From the distribution of ligand constituent, $\overline{C}_S(x)$ as a function of x, an experimental parameter, $\Omega_S(x)$, is defined operationally as

$$\Omega_S(x) = [\overline{C}_S(x)/\overline{C}_S(x_F)] \exp [M_S(1 - \bar{v}_S\rho)\omega^2(x_F^2 - x^2)/2\mathbf{RT}] \quad (4.11)$$

For $[\overline{C}_S(x), x]$ points throughout the distribution, one of those points being taken as $[\overline{C}_S(x_F), x_F]$. The resulting values of $\Omega_S(x)$ are then plotted as a function of $\overline{C}_S(x)$ for extrapolation to zero ligand concentration to obtain $(\Omega_S)_o$. On the grounds that

$$(\Omega_S)_o = \lim_{\overline{C}_S(x) \to 0} \Omega_S(x) = a_S(x_F)/\overline{C}_S(x_F) \quad (4.12)$$

this ordinate intercept of the $\Omega_S(x)$ plot defines the thermodynamic activity of free ligand at the chosen reference radial position, x_F. Incorporation of this value of $a_S(x_F) = (\Omega_S)_o\overline{C}_S(x_F)$ into eq. 4.10 then gives $a_S(x)$ for all experimental points of the sedimentation equilibrium distribution. As in all previous methods, thermodynamic ideality is assumed, whereupon the binding function, $r(x)$, at all points throughout the distribution, may be obtained as

$$r(x) = [\overline{C}_S(x) - a_S(x)]/\overline{C}_A(x) \quad (4.13)$$

in which the magnitude of $\overline{C}_A(x)$ is obtained from the separate sedimentation equilibrium distribution for acceptor component. In principle, a range of $[r(x), C_S(x)]$ points is thus obtained from a single sedimentation equilibrium experiment. However, if S were relatively small, C_S would be essentially constant throughout the distribution (eq. 4.10), and hence multiple estimates of a single (r, C_S) point would be obtained for the binding curve. Under those circumstances it is simpler to determine r and C_S as the slope and ordinate intercept, respectively, of a plot of $\overline{C}_S(x)$ *versus* $\overline{C}_A(x)$, as discussed by Steinberg and Schachman (1966) in relation to sedimentation equilibrium studies of mixtures of bovine serum albumin and methyl orange. That system is, of course, amenable to quantitative study by the far simpler procedures of equilibrium dialysis and gel chromatography (Chapter 2).

The ability to obtain separate distributions for the acceptor and ligand constituents is a desirable but not essential prerequisite for application of the omega analysis to sedimentation data reflecting interactions between dissimilar proteins (Nichol *et al.*, 1976; Jeffrey *et al.*, 1979). However, not surprisingly, the analysis is more complicated in experiments where the sole equilibrium distribution available is a refractometric record. The only concentration that may be determined is the combined weight concentration of both constituents, $\bar{c}(x) = \bar{c}_A(x) + \bar{c}_S(x)$, a quantity obtained on the basis that all proteins exhibit very similar specific refractive increments.

Under these circumstances an $\Omega_S(x)$ function may still be determined from the weight concentration distribution, $\bar{c}(x)$ *versus* x, as

$$\Omega_S(x) = [\bar{c}(x)/\bar{c}(x_F)] \exp [M_S(1 - \bar{v}_s\rho)\omega^2(r_F^2 - r^2)/2\mathbf{RT}] \quad (4.14)$$

and extrapolated to zero solute concentration $[\bar{c}(x) \rightarrow 0]$ to obtain $(\Omega_S)_o$ and hence $a_S = (\Omega_S)_o\bar{c}(x_F)$, the procedure adopted above (eq. 4.12). As before, eq. 4.10 may now be used to define the thermodynamic activity of free ligand throughout the distribution, the only difference being that $a_S(x) \approx c_S(x)$ is a weight concentration in this case. Furthermore, the lack of information about the separate distribution of total ligand means that there is no analog to eq. 4.13 for the calculation of $r(x)$.

In order to proceed with the analysis the calculated concentrations of free ligand, $c_S(x)$, are subtracted from the corresponding values of $\bar{c}(x)$ to generate an amended sedimentation equilibrium distribution, $\bar{c}^*(x)$ *versus* x, in which acceptor (A) is now the smallest macromolecular species. A second omega function, $\Omega_A^*(x)$, may therefore be calculated as

$$\Omega_A^*(x) = [\bar{c}^*(x)/\bar{c}^*(x_F)] \exp [M_A(1 - \bar{v}_A\rho)\omega^2(x_F^2 - x^2)/2\mathbf{RT}] \quad (4.15)$$

and again extrapolated to $\bar{c}^*(x) = 0$ to obtain the ordinate intercept, $(\Omega_A^*)_o = a_A(x_F)/\bar{c}^*(x_F)$, whereupon the weight concentration of acceptor may also be calculated throughout the distribution. The distribution of combined complex species (in g/liter) may then be obtained as $\bar{c}(x) - c_S(x) - c_A(x)$.

Figure 4.5 illustrates this double-omega analysis of the Rayleigh interferometric record of a sedimentation equilibrium distribution for a mixture of ovalbumin (A) and lysozyme (S). For that particular system the subsequent analysis is complicated by a reaction stoichiometry greater than unity (Jeffrey *et al.*, 1979); but we shall, for simplicity, only describe the procedure to be followed for single complex formation. The difficulty arises from the fact that the distributions of free S, free A, and combined AS_i species are evaluated in terms of weight concentrations, whereas determination of the binding constant depends upon expression of these concentrations in molar terms. Provided that complex formation is restricted to AS, this requirement poses no problem because the molar concentrations of all three species, $C_S(x) = c_S(x)/M_S$, $C_A(x) = c_A(x)/M_A$, and $C_{AS}(x) = [\bar{c}(x) - c_S(x) - c_A(x)]/M_{AS}$, may be determined. Readers who wish to proceed to analysis of the more general case with $p > 1$ are referred to the original article on the ovalbumin–lysozyme system (Jeffrey *et al.*, 1979).

4.4. ISOTHERMAL TITRATION CALORIMETRY

The uptake or release of heat is a universal property of chemical reactions with non-zero enthalpy change. For an acceptor–ligand interaction, the extent of this heat uptake or release is proportional to the amount of ligand bound and hence provides a means of monitoring the course of the reaction. Isothermal

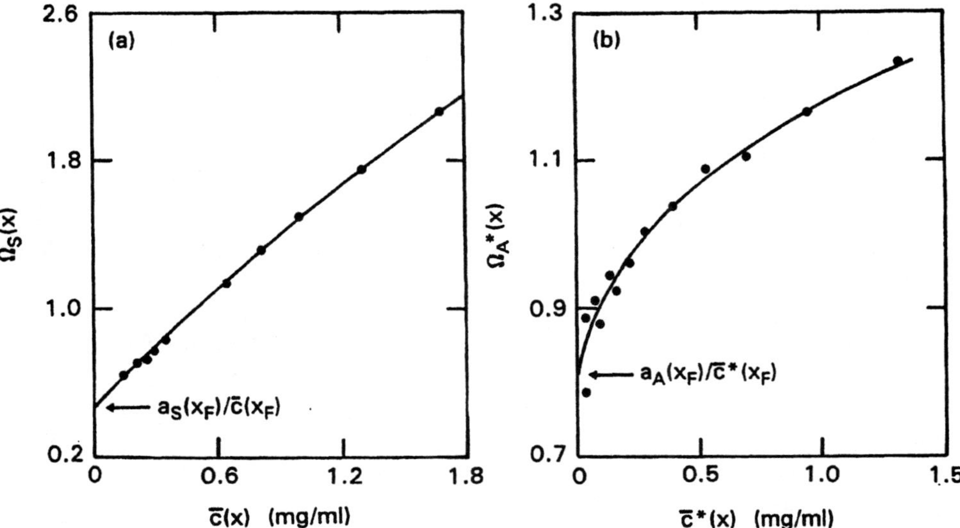

Fig. 4.5. Determination of the free concentrations of lysozyme (S) and ovalbumin (A) by omega analysis of a sedimentation equilibrium distribution for an interacting mixture of the two proteins. **(a)** Plot of the dependence of $\Omega_S(x)$, defined by eq. 4.14, upon total concentration $\bar{c}(x)$ to obtain the thermodynamic activity of lysozyme, $a_S(x_F)$, in the reference mixture with $\bar{c}(x_F) = 0.52$ mg/ml. **(b)** Plot of the corresponding dependence of $\Omega_A^*(x)$, calculated according to eq. 4.15, upon $\bar{c}^*(x)$ to obtain the thermodynamic activity of A at the same reference position. [Adapted from Jeffrey *et al.* (1979).]

microcalorimeters measure the temperature difference between a sample cell containing the acceptor to which ligand is added and a reference cell containing the acceptor only, and indicate the amount of heat required to keep the temperature difference either zero or constant (Freire *et al.*, 1990). If the acceptor possesses only a single site for interaction with ligand, the change in the concentration of bound ligand is related to the heat change $dQ \approx \delta Q$ by the expression (Wiseman *et al.*, 1989)

$$d_Q = d(\overline{C}_S - C_S)V\Delta\mathbf{H}^\circ \tag{4.16}$$

where $\Delta\mathbf{H}^\circ$ is the molar enthalpy of binding and V is the cell volume. From the definition of the association equilibrium constant, it follows that the concentration of bound ligand, namely,

$$(\overline{C}_S - C_S) = (1/2)\{(\overline{C}_A + \overline{C}_S + 1/k_{AS}) - [(\overline{C}_A + \overline{C}_S + 1/k_{AS})^2 - 4\overline{C}_A\overline{C}_S]^{1/2}\} \tag{4.17}$$

Differentiating eq. 4.17 with respect to $\overline{C}_S$ and substituting the consequent expression into eq. 4.16 give (Wiseman *et al.*, 1989)

$$(1/V)(dQ/d\overline{C}_S)$$
$$= \Delta\mathbf{H}^\circ\left[\frac{1}{2} + \frac{1 - [1 + 1/(k_{AS}\overline{C}_A)]/2 - \zeta/2}{\{\zeta^2 - 2\zeta[1 - 1/(k_{AS}\overline{C}_A)] + [1 + 1/(k_{AS}\overline{C}_A)]^2\}^{1/2}}\right] \tag{4.18}$$

which shows that the differential heat, $dQ/d\overline{C}_S$, is dependent on $\zeta = \overline{C}_S/\overline{C}_A$, the molar ratio of reactants, and $k_{AS}\overline{C}_A$, the product of the association constant and the total acceptor concentration. Nonlinear regression analysis of the dependence of $(1/V)(dQ/d\overline{C}_S)$ upon $\overline{C}_S/\overline{C}_A$ in terms of eq. 4.18 is used to obtain k_{AS} and $\Delta\mathbf{H}^\circ$. In that regard the simulated plots presented in Figure 4.6 for a range of $k_{AS}\overline{C}_A$ values draws attention once again to the desirability of determining equilibrium constants under conditions where the total acceptor concentration is in the vicinity of the dissociation constant ($k_{AS}\overline{C}_A$ relatively small). Large values of $\overline{C}_A$ are thus required for the characterization of weak binding, whereas the limitation to studies of strong binding is not the low value of $\overline{C}_A$ but rather the sensitivity of the microcalorimeter and the magnitude of $\Delta\mathbf{H}^\circ$.

Experimental application of this technique is illustrated in Figure 4.7 with a calorimetric study of the interaction between a mannose disaccharide and concanavalin A (Williams *et al.*, 1992). Figure 4.7a depicts the experimental record obtained on adding successive aliquots of disaccharide to concanavalin A and thereby illustrates the diminution of the signal as saturation is approached. Analysis of the results in terms of eq. 4.18 requires evaluation of $Q/\overline{C}_S$ and hence integration of the peak areas to obtain the amount of heat liberated (Q). Figure 4.7b shows the dependence of $Q/(V\overline{C}_S)$ upon injection number, which is clearly analogous to $\overline{C}_S/\overline{C}_A$ (the suggested abscissa). Nonlin-

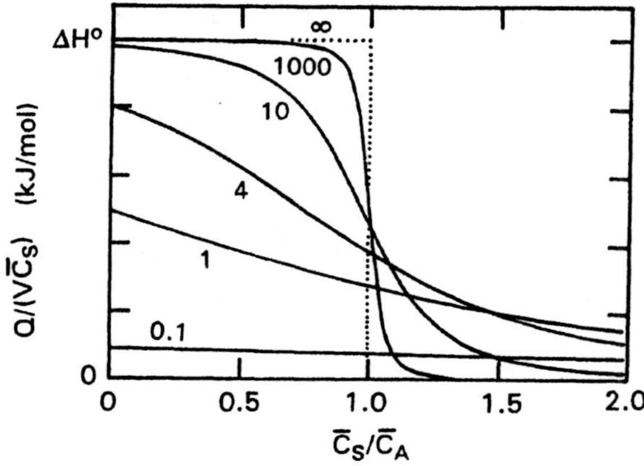

Fig. 4.6. Theoretical binding isotherms illustrating the range of behavior likely to be encountered in studies of acceptor–ligand interactions by microcalorimetry. The number adjacent to each curve denotes the value of $k_{AS}\overline{C}_A$ used in eq. 4.18 for its simulation.

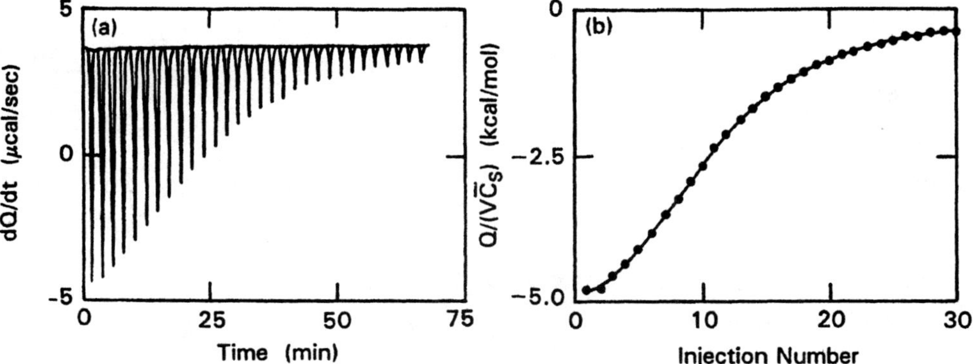

Fig. 4.7. Isothermal titration of concanavalin *A* (0.4 mM initially) with methyl 6-*O*-(α-D-mannopyranosyl)-α-D-mannopyranoside. **(a)** Experimental trace showing the rate of heat generation for 30 successive injections of disaccharide (22 mM). **(b)** Corresponding plot of total heat Q (obtained by integrating the areas in Fig. 4.6a) that is used for determining k_{AS} and ΔH°. [Adapted from Williams *et al.* (1992).]

ear regression analysis in terms of the counterpart of eq. 4.18 yields values of 15,000 M^{-1} and −7.8 kcal/mol for k_{AS} and **ΔH**°, respectively. By calculating the standard free energy change (**ΔG**° = −**RT** ln k_{AS}), the standard enthropy change may also be determined as **ΔS**° = (**ΔH**° − **ΔG**°)/**T**.

Apart from the widespread applicability of the technique, microcalorimetry clearly has an advantage over other methods in that values for k_{AS}, the number of binding sites, the standard enthalpy change (**ΔH**°), the standard free energy change (**ΔG**°), and the standard entropy change (**ΔS**°) can all be obtained from the one experiment. Furthermore, the above analysis is also readily extended to cover situations in which the acceptor exhibits more than one class of binding site for ligand (Lin *et al.*, 1991). In addition, the technique can be used to examine the binding of ligands to particulate structures (*e.g.*, ligand interactions with membrane receptors) and to an unpurified acceptor provided that the additional components are inert and do not perturb the acceptor–ligand interaction.

This section has demonstrated the manner in which ligand binding can be characterized quantitatively by isothermal titration microcalorimetry, a technique that seems to be gaining widespread popularity as the calorimetric method of choice for this purpose (Freire *et al.*, 1990). The other principal technique, differential scanning calorimetry, entails measurement of the heat capacity (temperature derivative of enthalpy) by subjecting the system to a change in temperature. It has been used mainly for investigating the influence of ligand binding on the stability of proteins against thermal denaturation (Privalov and Khechinashvili, 1974). However, the feasibility of using differential scanning calorimetry to quantify the energetics of ligand binding has also been demonstrated (Robert *et al.*, 1988, 1989; Brandts and Lin, 1990).

5

BINDING CONSTANTS FROM COMPETITIVE BINDING ASSAYS

Whereas the preceding three chapters have considered the problem of evaluating an equilibrium constant by direct study of an interaction, emphasis is now turned to the use of indirect procedures whereby equilibrium parameters for the interaction in question are inferred from the inhibitory effect of a competing ligand (I) on an acceptor–ligand interaction that has already been characterized. Examples of indirect characterization are to be found in the screening of drug or antigen analogs by radioimmunoassay and ELISA techniques and also in the study of interactions by quantitative affinity chromatography. Because the theory of such competitive binding assays has been considered most extensively in the context of quantitative affinity chromatography, we begin this chapter on that note.

5.1. QUANTITATIVE AFFINITY CHROMATOGRAPHY

Quantitative affinity chromatography (Andrews *et al.*, 1973; Dunn and Chaiken, 1974; Nichol *et al.*, 1974b) is arguably becoming the most versatile of the methods available for evaluating ligand-binding constants (Winzor, 1992; Winzor and Jackson, 1993), this being an assertion (Winzor, 1985) that stems from the fact that the experimental parameter measured is the distribution of partitioning solute (acceptor, A) between soluble and matrix-bound states. Consequently, the major requirement for its application is an ability to measure the position of this equilibrium, irrespective of the ligand concentration required to effect the particular equilibrium distribution of solute (Bergman and Winzor, 1986; Waltham *et al.*, 1988; Hogg *et al.*, 1991; Winzor *et al.*, 1992). Envisaged initially as a method of characterizing interactions that are too weak for study by conventional means (Nichol *et al.*, 1974b, 1981b), quantitative affinity chromatography has, in recent years, been adapted for the characterization of interactions at the other end of the energy spectrum, namely, high-affinity interactions that are too strong for study by conventional means (Waltham *et al.*, 1988; Hogg *et al.*, 1991; Ward *et al.*, 1995).

5.1.1. General Theoretical Aspects

The first step in quantitative affinity chromnatography is to choose a matrix that, either by virtue of its intrinsic chemical properties or as the result of covalent attachment of a specific reactant, interacts specifically with particular molecules in solution. The matrix-associated reactant (X) is thus chosen to obtain selective interaction with the desired solute of interest (A). For example, the interaction of antithrombin (A) with heparin (I) has been studied (Hogg *et al.*, 1991) by using heparin immobilized on Sepharose as the matrix-bound reactant (X), whereas unmodified Sephadex G-50 is already an adequate affinity matrix for studies of the interaction between lectins and saccharides (Hogg and Winzor, 1987a; Hogg *et al.*, 1987b) or glycoproteins (Hogg and Winzor, 1987a). For systems with only two types of interaction there are two potential competitive situations that can arise. Inhibitor-facilitated elution of acceptor occurs if the interactions of solute (A) with matrix sites (X) and inhibitor (I) are mutually exclusive (Fig. 5.1a) or if the competition is between solute and inhibitor for matrix sites (Fig. 5.1b).

The first step in a quantitative affinity chromatography study entails determination of the binding constant for the solute–matrix interaction, k_{AX}, and an associated parameter, $\bar{\bar{C}}_X$, the effective total concentration of affinity matrix sites. Provided that the partitioning solute is univalent in its interaction with matrix sites, the binding function, r, may be defined in the usual way (*cf.* eq. 1.3) as

$$r = (\bar{\bar{C}}_A - \bar{C}_A)/\bar{\bar{C}}_X \tag{5.1}$$

where $\bar{C}_A$ denotes the concentration of A in the liquid phase and $\bar{\bar{C}}_A$, $\bar{\bar{C}}_X$ the respective total concentrations of solute and matrix sites in the whole system.

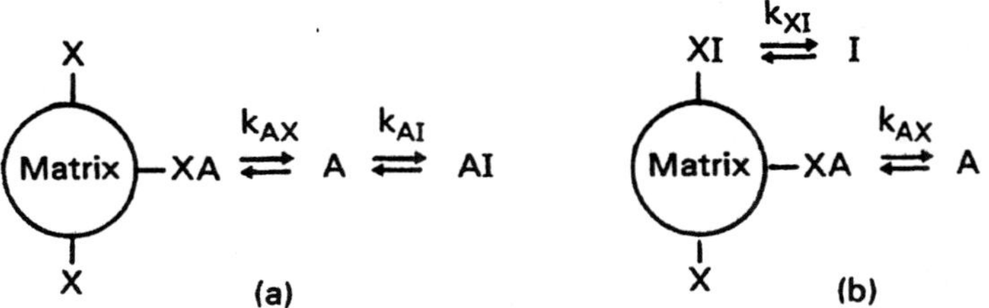

Fig. 5.1. Schematic representation of the two competitive situations encountered in the characterization of ligand binding by quantitative affinity chromatography.

Because $\bar{\bar{C}}_X$ is one of the parameters to be evaluated from the analysis, eq. 5.1 is used experimentally to define the product $r\bar{\bar{C}}_X$ as the difference between $\bar{\bar{C}}_A$ and $\bar{C}_A$. The value of $(\bar{\bar{C}}_A - \bar{C}_A)$ is then incorporated into the Scatchard (1949) linear transform of the rectangular hyperbolic expression for solute binding to a single class of matrix sites with intrinsic affinity constant k_{AX} (eq. 1.10a) to give

$$(\bar{\bar{C}}_A - \bar{C}_A)/\bar{C}_A = k_{AX}\bar{\bar{C}}_X - k_{AX}(\bar{\bar{C}}_A - \bar{C}_A) \quad (5.2)$$

By conducting a series of experiments on mixtures with a fixed concentration of free competing ligand, C_I, eq. 5.2 may still be employed to describe the partitioning of solute in terms of an effective association constant, $\bar{k}_{AX}$, where the overbar notation is used to denote a constitutive quantity (Baghurst and Nichol, 1975; Kuter *et al.*, 1983; Winzor, 1992; Winzor and Jackson, 1993). In that regard the partition behavior of the solute clearly requires description in terms of two phenomena—the solute–matrix interaction governed by k_{AX} and either a solute–inhibitor interaction governed by binding constant k_{AI} or a matrix–inhibitor interaction governed by k_{XI} (Fig. 5.1). For the former situation (Fig. 5.1a) the value of $\bar{k}_{AX}$ obtained by applying eq. 5.2 to results obtained in the presence of ligand is being calculated on the basis that $\bar{C}_A$ (the concentration of solute in the liquid phase) is the free concentration, whereas the actual concentration of free solute, C_A, is smaller than $\bar{C}_A$ because of solute–ligand complex formation. The two concentrations are related by the expression $C_A = \bar{C}_A/(1 + k_{AI}C_I)$, where C_I is the free inhibitor concentration. It then follows that the two solute–matrix affinity constants are related by

$$k_{AX} = \bar{k}_{AX}(1 + k_{AI}C_I) \quad (5.3a)$$

k_{AI} may thus be evaluated from the slope of the linear dependence of $k_{AX}/\bar{k}_{AX}$ upon C_I (Kuter *et al.*, 1983; Winzor, 1992; Winzor and Jackson, 1993). An analogous expression, namely,

$$k_{AX} = \bar{k}_{AX}(1 + k_{XI}C_I) \quad (5.3b)$$

is obtained for the other competitive case (Fig. 5.1b) because $\bar{k}_{AX}$ is obtained on the basis that the total concentration of matrix sites accessible to A is $\bar{\bar{C}}_X$ rather than $\bar{\bar{C}}_X/(1 + k_{XI}C_I)$. Knowledge of $\bar{C}_I$, the total inhibitor concentration (rather than C_I) for systems with a competitive solute–ligand interaction, is accommodated by noting (Hogg *et al.*, 1991; Winzor *et al.*, 1992) that C_I is given by

$$C_I = \bar{C}_I - (Q - 1)\bar{C}_A/Q \quad (5.4)$$

whereupon k_{AI} is conveniently evaluated from the relationship

$$Q = k_{AX}/\bar{k}_{AX} = 1 + k_{AI}[\bar{C}_I - (Q - 1)\bar{C}_A/Q] \quad (5.5a)$$

For systems in which the competitive interaction of ligand is with matrix sites, k_{XI} may be calculated from the expression (Winzor and Jackson, 1993)

$$(Q - 1)k_{AX}\overline{C}_A = k_{XI}[k_{AX}\overline{C}_A\overline{\overline{C}}_I - (Q - 1)(\overline{\overline{C}}_A - \overline{C}_A)] \qquad (5.5b)$$

where $\overline{\overline{C}}_I$ (rather than $\overline{C}_I$) is used to denote that the total inhibitor concentration inferred from the amount added and the accessible liquid-phase volume includes a contribution from inhibitor attached to the matrix phase.

The next question to be addressed concerns the experimental protocol to be used for the quantitative affinity chromatography study. Unless the experimenter has access to biosensor technology (see Section 5.2), there are two options from which to choose—column chromatography and partition equilibrium studies. These are considered in turn, together with the means of evaluating the concentration difference required for the application of eqs. 5.2 and 5.3.

5.1.2. Partition Equilibrium Studies

The most direct method of employing eqs. 5.2 and 5.3 is to perform simple partition experiments on matrix-solute mixtures (with and without inhibitor) and to measure the matrix-bound concentration of solute, $(\overline{\overline{C}}_A - \overline{C}_A)$, as the difference between the concentration inferred from the amount of solute added to the slurry $(\overline{\overline{C}}_A)$ and $\overline{C}_A$, the concentration of A measured in the liquid phase after attainment of equilibrium. This simple procedure and its recycling partition counterpart have been described in Section 2.3.2 as methods of measuring free ligand concentrations in mixtures containing a particulate acceptor. Although the basic experimental aspects remain unchanged from those described in that context, the presence of a third reactant (competing ligand I) is an additional complexity in the use of the partition equilibrium method as a competitive binding assay. In the recycling partition technique (see Fig. 2.6a) the solute–matrix interaction is thus characterized by the procedure described in Section 2.3.2, that is, by adding aliquots of a stock solute solution of partitioning solute to the stirred slurry and determining the resulting value of $\overline{C}_A$ after each addition from the equilibrium monitor response. Successive aliquots of a stock solution of the competing ligand (I) are then added to provide the corresponding $[\overline{\overline{C}}_A, \overline{C}_A, \overline{C}_I]$ results for the evaluation of $\overline{k}_{AX}$ as a function of $\overline{C}_I$ (or $\overline{\overline{C}}_I$).

Application of the recycling procedure is illustrated by considering the results of an investigation with heparin-Sepharose as the affinity matrix for characterization of the interaction between antithrombin and high-affinity heparin (Hogg *et al.*, 1991). In the evaluation of k_{AX} for the interaction of antithrombin with the heparin-Sepharose affinity matrix via eq. 5.2, allowance needs to be made for the decrease in $\overline{\overline{C}}_X$ as the result of the increases in the liquid-phase volume (V_A^*). In keeping with the procedure adopted in Section 2.3.2, this dilution is taken into account by multiplying the conventional ordinate and abscissa parameters of the Scatchard plot by $V_A^*/(V_A^*)_0$, such plots being shown in Figure 5.2a for the present system at three temperatures. The effective total concentration of matrix sites that is defined by the abscissa intercept is $(\bar{\bar{C}}_X)_0$, the value at the commencement of solute addition. The common abscissa intercept in Figure 5.2a signifies a magnitude of 2.2 μM for $(\bar{\bar{C}}_X)_0$; and the slopes yield values of 13.0 ($\pm$1.1), 6.8 ($\pm$0.3), and 3.0 ($\pm$0.3) $\times$ 10^6 M^{-1} for k_{AX} at 15°, 25°, and 35°C.

Determination of $(\bar{\bar{C}}_X)_0$ allows values of $\bar{\bar{C}}_X$ to be calculated for each stage of the second part of the experiment involving the displacement of antithrombin from the affinity matrix by soluble heparin (I). This allows $\bar{k}_{AX}$ to be calculated as the only parameter of unknown magnitude in eq. 5.2. Furthermore, because $\bar{C}_I$, the total concentration of high-affinity heparin, is the quantity obtained by dividing the amount added by V_A^*, results need to be plotted in accordance with eq. 5.5a to obtain k_{AI} (Fig. 5.2b). The slopes indicate binding constants of 8.0 ($\pm$2.2), 3.4 ($\pm$0.3), and 1.0 ($\pm$0.2) $\times$ 10^7 M^{-1} for the interactions of antithrombin with high-affinity heparin at the three temperatures. A value of -4.2 kcal/mol is obtained for the standard enthalpy change (ΔH°) from the slope ($-\Delta$H°/R) of the dependence of ln k_{AI} upon 1/T (a plot according to the integrated form of eq. 1.1).

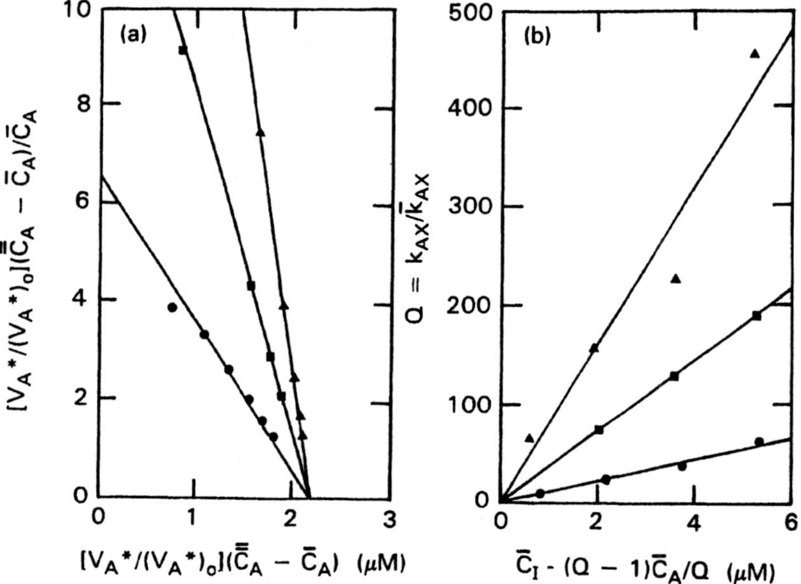

Fig. 5.2. Characterization of ligand binding by the recycling partition technique of quantitative affinity chromatography. (**a**) Scatchard plot of results [from Table 2 of Hogg *et al.* (1991)] for the interactions of antithrombin with heparin–Sepharose at 15° (▲), 25° (■), and 35°C (●). (**b**) Evaluation of the binding constant (k_{AI}) for the corresponding interactions of antithrombin (A) with high-affinity heparin (I) from results (also presented in the abovementioned table) plotted in accordance with eq. 5.5a. [Adapted from Hogg *et al.*, (1991).]

5.1.3. Column Chromatographic Procedures

In the adaptation of the above binding expressions to column chromatography, advantage is taken of the fact that the total amount of solute, $V_A^* \overline{\overline{C}}_A$, must equal $\overline{V}_A \overline{C}_A$, the product of the concentration of A in the liquid phase and the measured elution volume, $\overline{V}_A$ (Hogg and Winzor, 1984). Upon substitution of $\overline{V}_A / V_A^*$ for $\overline{\overline{C}}_A / \overline{C}_A$, eq. 5.2 becomes

$$[(\overline{V}_A / V_A^*) - 1] = k_{AX} \overline{\overline{C}}_X - k_{AX} \overline{C}_A [(\overline{V}_A / V_A^*) - 1] \qquad (5.6)$$

which is the column chromatographic equivalent of the Scatchard plot. Clearly, the evaluation of k_{AX} (or $\overline{k}_{AX}$) from such a plot requires knowledge of $\overline{C}_A$, the total concentration of solute in the liquid phase. Because this quantity is not available from conventional zonal chromatography experiments, rigorous application of eq. 5.5 requires use of frontal chromatography (see Section 2.4.1). In that regard the use of small columns can decrease substantially the volume of solution that needs to be loaded onto the column in order to generate an elution profile that contains a plateau region in which the concentrations of all species (A, I, and AI if relevant) equal those in the applied solutions (Kyprianou and Yon, 1982). In frontal chromatography the required elution volume ($\overline{V}_A$) is given by the median bisector of the boundary in solute constituent (Winzor and Jackson, 1993). For a reasonably symmetrical boundary, $\overline{V}_A$ may be approximated by the effluent volume at which $\overline{C}_A$ attains half of its value in the plateau region.

Application of this procedure is illustrated in Figure 5.3 by results from an investigation in which the competition between *p*-nitrophenylmannoside (A) and methylmannoside (I) for immobilized lectin sites (X) has been characterized by frontal chromatography with an affinity matrix comprising concanavalin A attached covalently to porous glass beads (Munro *et al.*, 1993). Open symbols denote the dependence of $\overline{V}_A$ upon $\overline{C}_A$, plotted according to eq. 5.6, obtained from experiments with *p*-nitrophenylmannoside alone, whereas the closed symbols describe the corresponding dependence from experiments in which the pre-equilibrating buffer and also the applied solute solution were supplemented with 100 μM methylmannoside ($\overline{\overline{C}}_I \approx C_I$). A binding constant ($k_{AX}$) of 27,000 M^{-1} for the interaction of *p*-nitrophenylmannoside is obtained from the slope of the plot in the absence of competing ligand. Combination of the corresponding value of 14,000 M^{-1} for $\overline{k}_{AX}$ with C_I = 100 μM in eq. 5.3b then yields a binding constant (k_{XI}) of 9,000 M^{-1} for the competitive interaction of methylmannoside with immobilized concanavalin A (Munro *et al.*, 1993).

Zonal affinity chromatography has been used extensively for characterizing the interaction between an enzyme and an inhibitor by the effect of I on the elution of the enzyme from a matrix with covalently bound inhibitor as immobilized reactant, X (Dunn and Chaiken, 1974). In zonal chromatography, however, the inability to ascribe a unique value to $\overline{C}_A$ because of the ever-changing solute concentration in the migrating zone clearly precludes appli-

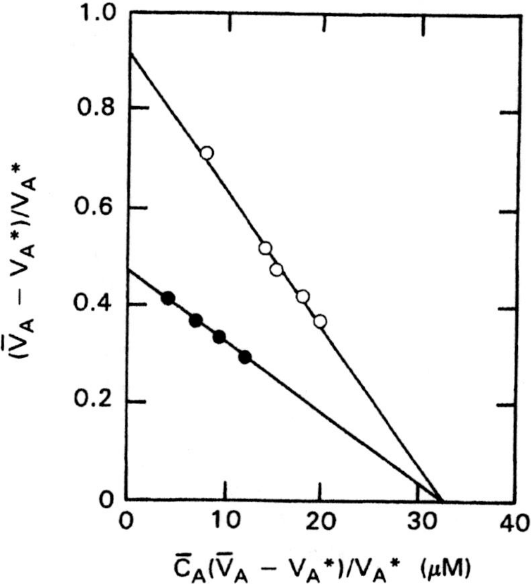

Fig. 5.3. Scatchard plots of results obtained in frontal affinity chromatography of *p*-nitrophenylmannoside (*A*) on a column of glyceryl-CPG 3000 with concanavalin *A* as immobilized ligand (*X*) in the absence of (○) and presence (●) of 100 μM methylmannoside (*I*). [Adapted from Munro *et al.* (1993).]

cation of the above rigorous procedure. Under those circumstances the only course of action is to truncate eq. 5.6 as

$$[(\bar{V}_A/V_A^*) - 1] = k_{AX}\bar{\bar{C}}_X \tag{5.7}$$

which entails the assumption that the total concentration of matrix sites, $\bar{\bar{C}}_X$, provides a sufficiently precise estimate of their free concentration, C_X (Bergman and Winzor, 1986). With that proviso, which applies if $\bar{\bar{C}}_X >> \bar{\bar{C}}_A$, zonal affinity chromatography of the solute on an affinity column pre-equilibrated with a range of inhibitor concentrations, C_I, allows the solute–ligand binding constant to be evaluated from the expression

$$[(\bar{V}_A/V_A^*) - 1] = k_{AX}\bar{\bar{C}}_X/(1 + k_{AI}C_I) \tag{5.8}$$

of which two linear transforms are

$$1/[(\bar{V}_A/V_A^*) - 1] = (1/k_{AX}\bar{\bar{C}}_X) + k_{AI}C_I/(k_{AX}\bar{\bar{C}}_X) \tag{5.9a}$$

and

$$[(\bar{V}_A/V_A^*) - 1] = k_{AX}\bar{\bar{C}}_X - k_{AI}C_I[\bar{V}_A/V_A^*) - 1] \tag{5.9b}$$

Although the former transform has traditionally been used for the analysis of zonal affinity chromatographic data (Dunn and Chaiken, 1974, 1975; Swaisgood and Chaiken, 1987), its disadvantage is that k_{AI} must be evaluated as the ratio of the slope to the ordinate intercept. Consequently, the use of a highly substituted affinity matrix ($\overline{\overline{C}}_X$ very large) may well give rise to an ordinate intercept ($1/k_{AX}\overline{\overline{C}}_X$) that is indistinguishable from zero. This situation was encountered in an attempt (Dunn *et al.*, 1983) to characterize the interaction of *p*-aminobenzamidine with trypsin by its effect on zonal affinity chromatography of the enzyme on *p*-aminobenzamidine–Sepharose (Fig. 5.4a). Fortunately, the consequent inability to evaluate k_{AI} is readily avoided by using eq. 5.9b (Bergman and Winzor, 1986), which allows a binding constant of 6,800 ($\pm$1,300) M^{-1} for the *p*-aminobenzamidine–trypsin interaction to be obtained directly from the slope (Fig. 5.4b). No separate estimate of either k_{AX} or $\overline{\overline{C}}_X$ emanates from zonal affinity chromatography, because these two parameters always appear as a product.

By rendering feasible the quantitative analysis of results obtained under conditions where the enforced consideration of $\overline{\overline{C}}_X$ to be the free concentration of matrix sites is most likely to be a valid approximation, eq. 5.9b thus achieves the breakthrough required for unequivocal characterization of ligand binding by zonal affinity chromatography. From the experimental viewpoint the zonal technique has the undeniable attraction of being more economical in terms of solute requirements. Furthermore, the fact that this affinity chromatographic counterpart of the Hummel–Dreyer gel chromatographic procedure (see Section

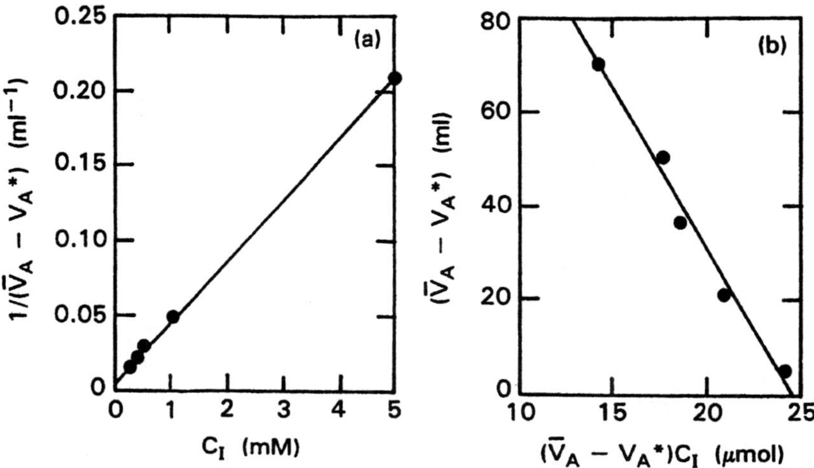

Fig. 5.4. Thermodynamic characterization of the interaction between *p*-aminobenzamidine (*I*) and trypsin (*A*) by zonal chromatography on *p*-aminobenzamidine–Sepharose. (**a**) Results in the form originally presented. (**b**) Their reassessment in terms of eq. 5.9b. [Adapted from Dunn *et al.* (1983) and Bergman and Winzor (1986), respectively.]

2.4.2) generates a solute zone migrating in the presence of a defined free concentration of ligand allows the technique to be used in studies of interactions where solute and ligand are both macromolecular (Hogg and Winzor, 1987a). Selection of a matrix that neither solute nor ligand can penetrate ensures satisfaction of the requirement, inherent throughout quantitative affinity chromatography theory, that a common accessible volume (V_A^*) applies to A and AI complex. Finally, the ability of zonal affinity chromatography to provide information on the elution volume ($\bar{V}_A$) in an environment with a known concentration of soluble ligand may also be used to advantage in studies with impure solute preparations, provided that an affinity matrix with absolute biospecificity for that solute is used (Brinkworth *et al.*, 1975).

5.2. USE OF BIOSENSOR TECHNOLOGY

An alternative way of quantifying solute partitioning between adsorbed and liquid states is to measure the concentration of bound solute in the equilibrium mixture by the recently developed biosensor technologies incorporated into the BIAcore or BIAlite (Pharmacia) and IASys (Fisons) instruments. An affinity chromatographic matrix is prepared by covalently attaching an appropriate reactant to a carboxymethylated dextran layer on the biosensor surface (a gold chip in the Pharmacia instruments, and the cuvet base in the Fisons instrument). Although different biosensor technologies are used in the two instruments (Löfås and Johnsson 1990; Cush *et al.*, 1993), they both exhibit similar sensitivities of response to bound solute. Because the mass of bound solute is the parameter being monitored, the methodology is far more sensitive, in molar terms, for macromolecular than small partitioning solutes. The optimal range of bound concentration is 5–100 nM for proteins. Originally, the major interest in biosensor-based instruments was their use to define the kinetics of adsorption and desorption for the solute–matrix interaction (Karlsson *et al.*, 1991; Zeder-Lutz *et al.*, 1993; O'Shannessy *et al.*, 1993), but attention has recently been drawn to their application in the context of quantitative affinity chromatography (Ward *et al.*, 1995; Kalinin *et al.*, 1995).

In the BIAcore and BIAlite instruments, which differ only in regard to the degree of automation, the injected solution of partitioning solute flows through a capillary channel over the biosensor—a gold chip with an area of 1.1 mm^2: The dimensions of the channel are such that the volume of solution in the immediate vicinity of the biosensor surface is only 0.06 μl. Injection of a considerably larger volume of solute solution (50 μl) gives rise to a response of the form shown in Figure 5.5. Because the solute solution in contact with the biosensor is being replaced continually, the time-independent (plateau) response, R_p, is directly proportional to the concentration of bound solute, $(\bar{\bar{C}}_A - \bar{C}_A)$, in equilibrium with $\bar{C}_A$, the concentration of partitioning solute in the injected solution.

For a univalent partitioning solute the measurement of R_p as a function of

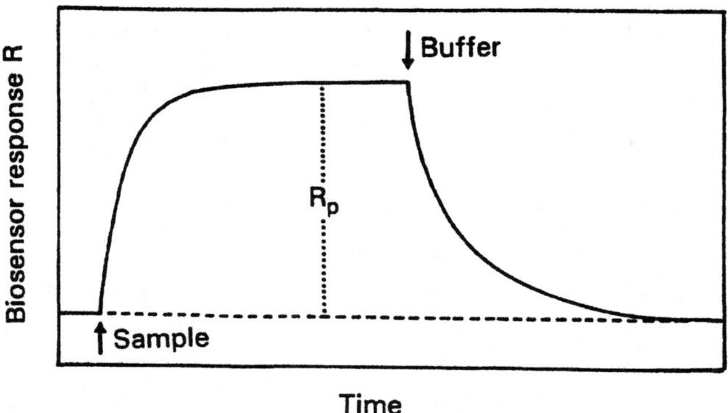

Fig. 5.5. Schematic representation of the time dependence of biosensor response resulting from injection of a relatively large volume (50 μl) of partitioning solute into the BIAcore (or BIAlite) instrument.

the injected solute concentration, $\overline{C}_A$, allows k_{AX} to be determined from the expression

$$R_p/\overline{C}_A = Bk_{AX}\overline{\overline{\overline{C}}}_X - k_{AX}R_p \tag{5.10}$$

a variant of the Scatchard formulation in which B is the proportionality constant relating the plateau response to bound concentration. The equilibrium constant for the solute–matrix interaction may thus be obtained from the slope of the linear dependence of $R_p/\overline{C}_A$ upon R_p. The equilibrium constant for the interaction between partitioning solute (A) and a competing ligand (I) is then obtained by injecting solute solution that has been supplemented with a range of known total concentrations of inhibitor, $\overline{C}_I$. Analysis of these data essentially follows the approach adopted in the recycling partition technique. Knowledge of the product $B\overline{\overline{\overline{C}}}_X$ (the abscissa intercept of the Scatchard plot) instead of $\overline{\overline{\overline{C}}}_X$ is accommodated by multiplying eq. 5.2 throughout by B, whereupon $\bar{k}_{AX}$ is determined from the expression

$$\bar{k}_{AX} = B(\overline{\overline{\overline{C}}}_A - \overline{C}_A)/\{\overline{C}_A[B\overline{\overline{\overline{C}}}_X - B(\overline{\overline{\overline{C}}}_A - \overline{C}_A)]\} \tag{5.11a}$$

$$= R_p/[\overline{C}_A(B\overline{\overline{\overline{C}}}_X - R_p)] \tag{5.11b}$$

Each experimental point, $[R_p, \overline{C}_A, \overline{C}_I]$, may thus be used to calculate the constitutive solute–matrix binding constant, $\bar{k}_{AX}$, whereupon k_{AS} may be evaluated by calculating $Q = k_{AX}/\bar{k}_{AX}$ for the required plot (eq. 5.5a) of Q *versus* $[\overline{C}_I - (Q - 1)\overline{C}_A/Q]$, as in Figure 5.2b. Use of the BIAcore instrument to obtain binding constants in this manner has been illustrated (Ward *et al.*, 1995) by attaching human interleukin-6 to the biosensor plate so that the soluble form of its specific receptor could be used as the partitioning solute (A) and human interleukin-6 as the competing ligand (I).

Detailed consideration of the use of the IAsys apparatus for rigorous thermodynamic characterization of ligand binding is premature in the sense that its use for such purposes is yet to be reported. The major difference between this instrumentation and the Pharmacia equipment is its use of a stirred cell rather than a flow-through system to establish equilibrium between partitioning solute in its soluble and biosensor-bound states. Consequently, although the time course of the biosensor response assumes the same form as that shown in Figure 5.5, the plateau response reflects a concentration of bound solute that is in equilibrium with a liquid-phase concentration, $\bar{C}_A$, which needs to be determined from $\bar{\bar{C}}_A$, the total concentration that is calculated from the amount of solute added, and the concentration of bound solute inferred from the biosensor response (Winzor, 1995). This need for conversion of R_p to a concentration of bound solute in order to define $\bar{C}_A$ (unless $\bar{\bar{C}}_A$ can justifiably be substituted for $\bar{C}_A$ because $(\bar{\bar{C}}_A - \bar{C}_A) \ll \bar{\bar{C}}_A$) is a disadvantage of the IAsys instrument; but the experimental design is clearly advantageous if partition equilibrium is attained very slowly. Furthermore, the method shares with the recycling partition variant of quantitative affinity chromatography the potential to provide complete characterization of the solute–matrix and solute–inhibitor interactions from a single experiment by successive additions of solute and inhibitor (see Fig. 5.3).

In summary, the use of biosensor technology to characterize ligand binding has not yet been explored sufficiently for any critical appraisal to be made of its potential advantages over other methods. Nevertheless, the results obtained thus far are sufficiently encouraging to indicate the likelihood that biosensor technology will play an important role in the thermodynamic characterization of biospecific interactions, particularly equilibria at the high end of the energy spectrum.

5.3. SCREENING OF LIGANDS BY RADIOIMMUNOASSAY

Although the screening of antigen analogs has traditionally entailed the quantitative analysis of competitive radioimmunoassays by the "logit" procedure (Rodbard *et al.*, 1968, 1969), such analyses are at best approximate because of nonfulfillment of the conflicting requirements that the concentration of radiolabeled antigen be both essentially zero and sufficiently large for saturation of all antibody sites (Hogg and Winzor, 1987b). The following procedure avoids this invalid assumption and also overcomes the shortcomings of the Müller (1980, 1983) method of radioimmunoassay.

5.3.1. Basis of the Procedure

As in any other competitive binding assay, the first step of a radioimmunoassay entails evaluation of the intrinsic binding constant, k_{AS}, for the interaction of ligand (antigen), S, with p-valent acceptor (antibody), A. By measuring the distribution of radiolabeled ligand between free and antibody-bound states in a series of mixtures with a fixed total antibody concentration, $\overline{C}_A$, and a range of total concentrations of antigen (radiolabeled and unlabeled), $\overline{C}_S$, this equilibrium constant is obtained on the basis that the antigen and its radiolabeled tracer form react indistinguishably. Under those circumstances, the Scatchard transform of the binding equation may be written

$$(1 - \alpha)/\alpha = pk_{AS}\overline{C}_A - k_{AS}(1 - \alpha)\overline{C}_S \qquad (5.12)$$

where $\alpha = C_S^*/\overline{C}_S^*$, the ratio of free to total concentrations of labeled antigen, is also considered to describe the corresponding distribution of unlabeled antigen.

In the second part of the experiment, reaction mixtures containing the same total concentrations of antibody, $\overline{C}_A$ and a fixed concentration of tracer-labeled antigen, $\overline{C}_S$, are supplemented with a range of total concentrations, $\overline{C}_I$, of antigen analog (inhibitor), I. The fraction of free labeled antigen in the presence of inhibitor, α_i, is now larger because some acceptor sites are occupied by the antigen analog. On the grounds that the concentration of free antigen sites is, from the definition of the intrinsic equilibrium constant, given by $(1 - \alpha_i)/(k_{AS}\alpha_i)$, the binding constant for the competing reaction, k_{AI}, may now be written

$$k_{AI} = (1 - \beta)\overline{C}_I k_{AS}\alpha_i/[(1 - \alpha_i)\beta\overline{C}_I] \qquad (5.13)$$

where β is the fraction of free inhibitor. The problem at hand is therefore to obtain an expression for $(1 - \beta)\overline{C}_I$, the concentration of antibody-bound inhibitor, as the difference between the total concentration of antibody sites and the combined concentrations of free and antigen-occupied antibody sites.

The total concentration of antibody sites may be calculated from the distribution of radiolabeled antigen in the absence of inhibitor as the sum of free and occupied antibody sites. Specifically, it can be seen from eq. 5.12 that

$$p\overline{C}_A = (1 - \alpha)/(k_{AS}\alpha) + (1 - \alpha)\overline{C}_S \qquad (5.14)$$

In a similar vein, the corresponding expression with α_i substituted for α also describes the combined concentrations of free and antigen-occupied antibody sites, whereupon the remainder are occupied by the competitive inhibitor. The concentration of bound inhibitor, $(1 - \beta)\overline{C}_I$, is therefore

$$(1 - \beta)\overline{C}_I = \overline{C}_S(\alpha_i - \alpha) - (1/k_{AS})[(1 - \alpha_i)/\alpha_i - (1 - \alpha)/\alpha] \qquad (5.15)$$

From the logarithmic form of eq. 5.13 it follows that a plot of log $\{(1 - \beta)\overline{C}_I[k_{AS}\alpha_i/(1 - \alpha_i)]\}$ *versus* log $(\beta\overline{C}_I)$ has a slope of unity and an ordinate intercept of log k_{AI}. Alternatively, the value of k_{AI} pertaining to the measurement of α_i for each $\overline{C}_I$ may be calculated by direct substitution of the values of $(1 - \beta)\overline{C}_I$ and $\beta\overline{C}_I$ into eq. 5.13.

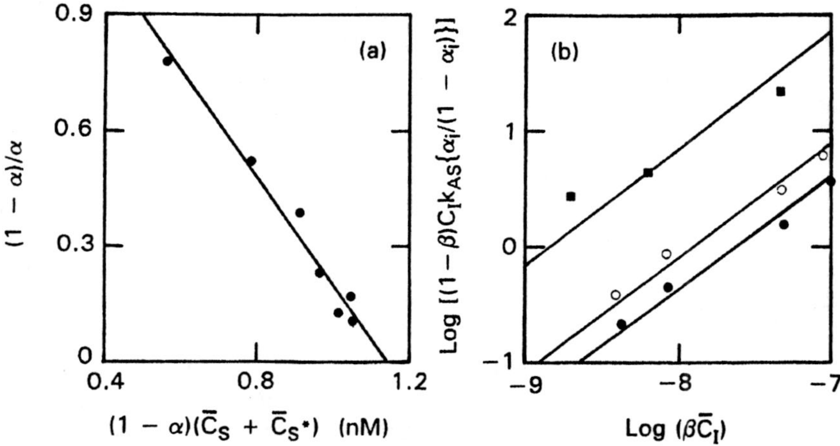

Fig. 5.6. Evaluation of the binding constant for an antibody–inhibitor interaction by quantitative radioimmunoaasay with a desipramine-specific antibody (Winzor *et al.*, 1991a). (**a**) Scatchard plot of results obtained in the absence of antigen analogs. (**b**) Plot of results obtained with mixtures of antibody and radiolabeled desipramine in the presence of imipramine (■), doxepin (○), and clomipramine (●) in accordance with the logarithmic form of eq. 5.13.

Such analysis of radioimmunoassay data is presented in Figure 5.6, which compares the interactions of a desipramine-elicited antibody with its antigen and several structural analogs thereof (Winzor *et al.*, 1991a). A value of 1.4 ($\pm$0.2) $\times$ 10^9 M^{-1} is obtained for k_{AS} from Scatchard analysis of the initial series of radioimmunoassays with unlabeled desipramine (Fig. 5.6a). Screening of other drugs is illustrated in Fig. 5.6b, which employs the logarithmic form of eq. 5.13 to display results for imiprimine, doxepin, and clomipramine. The finding that the monoclonal antibody binds to all three drugs with decreasing affinity (9 $\times$ 10^8, 9 $\times$ 10^7, and 4 $\times$ 10^7 M^{-1}) is consistent with their structural similarity to the eliciting antigen (Winzor et al., 1991a).

5.3.2. The Müller Method

At this stage it is appropriate to examine the Müller (1980, 1983) method for evaluating antibody–antigen affinity constants. Briefly, the procedure entails the conduct of a series of assays with a fixed antibody concentration and a range of labeled antigen concentrations ($\bar{C}_S^*$) to determine the concentration of radiolabeled antigen required for half to be bound in the absence of cold antigen. A second series of experiments is then performed on solutions containing that particular concentration of labeled antigen but supplemented with a range of concentrations of cold antigen ($\bar{C}_S$) to determine the concentration of the latter required to decrease the proportion of bound S^* by a further 50%. For the screening of an antigen analog (I), the cold antigen is replaced by the competitive inhibitor in another series of radioimmunoassays. To assess the merits of this approach, an alternative analysis needs to be adopted for the evaluation of k_{AS}.

For the radioimmunoassay conducted in the absence of cold antigen, the free (C_{S*}) and total ($\overline{C}_{S*}$) concentrations of labeled antigen are interrelated by the definition of the intrinsic binding constant (k_{AS*}) as

$$k_{AS*} = (1 - \alpha)/\{\alpha[p\overline{C}_A - \overline{C}_{S*}(1 - \alpha)]\} \qquad (5.16)$$

where α continues to define the free fraction of labeled antigen. Rearrangement of eq. 5.16 as

$$pk_{AS*}\overline{C}_A = (1 - \alpha)/\alpha + k_{AS*}\overline{C}_{S*}(1 - \alpha) \qquad (5.17)$$

gives an expression for the product of the binding constant and the total concentration of antibody sites ($p\overline{C}_A$). Provided that the antibody exhibits the same affinity for hot and cold antigen ($k_{AS*} = k_{AS}$), the corresponding expression for the experiments with mixtures of S and S^* is

$$pk_{AS*}\overline{C}_A = (1 - \alpha_s)/\alpha_s + k_{AS*}(\overline{C}_{S*} + \overline{C}_S)(1 - \alpha) \qquad (5.18)$$

where $\alpha_s = C_{S*}/\overline{C}_{S*}$ is the free fraction of antigen (labeled and cold) in the presence of total antigen concentration ($\overline{C}_{S*} + \overline{C}_S$). Because the product $pk_{AS*}\overline{C}_A$ is common to eqs. 5.17 and 5.18, they may be combined to give

$$(1 - \alpha)/\alpha - (1 - \alpha_s)/\alpha_s = k_{AS*}[(1 - \alpha_s)(\overline{C}_{S*} + \overline{C}_S) - (1 - \alpha)\overline{C}_{S*}] \qquad (5.19)$$

in which k_{AS*} is the only parameter of unknown magnitude. On making the substitutions $(1 - \alpha_s) = 0.5(1 - \alpha) = 0.25$ to accommodate the information extracted from the two series of radioimmunoassays, eq. 5.19 simplifies to

$$k_{AS*} = 8/[3(\overline{C}_S - \overline{C}_{S*})] \qquad (5.20)$$

which is eq. 13 of Müller (1980) in present terminology. That approach to obtain k_{AS*} is thus valid, but does not take full advantage of the available experimental data. Whereas eq. 5.20 only provides an estimate of k_{AS*} ($=k_{AS}$) based on empirically determined values of $\overline{C}_S$ and $\overline{C}_{S*}$ (to give 25% and 50% binding, respectively, of labeled ligand), eq. 5.19 allows all experimental data to be used collectively for the evaluation of the binding constant. This advantage also applies to eq. 5.12.

Although the Müller procedure provides a thermodynamically rigorous estimate of k_{AS} for a univalent antigen, the application of eq. 5.20 with $\overline{C}_S$ replaced by $\overline{C}_I$, the total concentration of a competitive inhibitor to effect a 50% reduction in the proportion of bound S^*, entails the assumption that $k_{AI} = k_{AS}$. This frequently used procedure is thus an invalid method of obtaining binding constants for a range of antigen analogs with different affinities for antibody. The basic tenets of the procedure may be retained, but for thermodynamic rigor the respective values of $\overline{C}_{S^*}$ and $\overline{C}_I$ pertaining to $\alpha = 0.5$ and $\alpha_i = 0.75$ need to be substituted into eq. 5.15. The result is

$$k_{AI} = k_{AS^*}(9k_{AS^*}\overline{C}_{S^*} + 24)/(12k_{AS^*}\overline{C}_I - 3k_{AS^*}\overline{C}_{S^*} - 8) \qquad (5.21)$$

which only simplifies to eq. 5.20 if $k_{AI} = k_{AS^*}$.

The application of eq. 5.21 requires the assignment of a magnitude to k_{AS^*}, which may be obtained from the series of experiments with cold antigen via eq. 5.19 or indeed from Scatchard analysis (eq. 5.12) of the initial series of assays in which $\overline{C}_{S^*}$ is varied to obtain the concentration required to achieve a value of 0.5 for α in the absence of cold antigen. In several respects the latter is the preferred procedure because it eliminates the need to assume that the chemical modification entailed in creating the radiolabeled antigen is without effect on the strength of its interaction with antibody. A detailed study of the interaction of ^{125}I-modified insulin with placental microsomal membranes (De Leo and Helmerhorst, 1992) has shown that iodination of one tyrosine (residue 14 of the α-chain) does yield a derivative that behaves indistinguishably from native insulin but that this requirement is not met by insulin iodinated at either of the other two tyrosine residues. One solution to that dilemma is to regard the cold antigen as a competitive inhibitor of the antibody interaction with radiolabeled ligand and therefore to obtain the required binding constant as k_{AI} via eq. 5.21 (or eq. 5.15). Adoption of this approach, which eliminates any assumption about the identity of k_{AS} (being determined as k_{AI}) and k_{AS^*}, does require the evaluation of k_{AS^*} from the initial series of radioimmunoassays with labeled antigen alone.

5.3.3. Antigen Multivalence and Other Complications

Although this treatment of the determination of binding constants by competitive radioimmunoassay serves to introduce the subject, it needs to be stressed that the quantitative expressions presented in this coverage refer only to situations in which the antigen is univalent. The adaptation of this approach to accommodate multivalence of the antigen has been illustrated in studies of the immunochemical reactivity of fibrinogen fragments (Winzor *et al.*, 1991b) and is discussed in Chapter 7. Consideration has also been given to situations in which the competition is only partial in the sense that the antibody exhibits additional sites for either the antigen or its analog (Brocklebank *et al.*, 1993).

5.4. QUANTIFICATION BY ELISA

Although solid-phase immunoassay procedures are used primarily for determining the concentrations of antigens in sera, essentially the same technique may be used to characterize the biospecific antibody–antigen interaction on which the immunoassay is based (Hogg and Winzor, 1987b; Hogg *et al.*, 1987a). In principle, expressions such as eqs. 5.2 and 5.5 could be used for analysis of solid-phase immunoassays (RIA or ELISA) in situations where the effect of soluble antigen (I) on the extent of partitioning of antibody (A) and immobilized antigen (X) allows accurate assessment of the difference between total antibody concentration ($\bar{\bar{C}}_A$) and $\bar{C}_A$, its constituent concentration in the liquid phase (Hogg and Winzor, 1987b). However, a problem with that approach in many immunoassays is that $\bar{\bar{C}}_X$ is so small in comparison with $\bar{\bar{C}}_A$ that there is no discernible difference between $\bar{\bar{C}}_A$ and $\bar{C}_A$. Under those circumstances it has become customary to remove the liquid phase and then determine the concentration of bound antibody, either by radioactivity measurements on the solid phase (RIA) or by means of an enzyme conjugated to an anti-immunoglobulin (ELISA). In resorting to this procedure it is imperative that the washing regimen be examined closely to establish (1) that it is adequate for the removal of all soluble antibody (free and complexed with soluble antigen I) and (2) that no discernible dissociation of matrix-bound antibody occurs during this washing regimen (Hogg and Winzor, 1987b; Hogg *et al.*, 1987a). In that regard the oft-used procedure of washing with cold buffer is not recommended because of the risk of altering the equilibrium position of any interaction with a non-zero standard enthalpy change ($\Delta H°$). Instead, the washing regimen should be varied by changing the number and/or volume of the washing steps performed at the temperature used for chemical equilibration.

The fact that $\bar{\bar{C}}_A$ greatly exceeds $\bar{\bar{C}}_X$ validates the substitution of the total antibody concentration ($\bar{\bar{C}}_A$) for its counterpart in the liquid phase ($\bar{C}_A$). Furthermore, the low concentration of immobilized antigen sites ($\bar{\bar{C}}_X$) usually justifies an additional approximation that complex formation between an antibody and immobilized antigen sites is effectively restricted to 1:1 stoichiometry despite multivalence of the antibody (Nichol *et al.*, 1974b; Hogg and Winzor, 1987b; Hogg *et al.*, 1987a). These approximations allow the experimentally determined ratio of the concentrations of antibody bound to a fixed concentration $\bar{\bar{C}}_X$ of matrix sites in the absence and presence of a concentration C_I of soluble antigen to be analyzed in terms of the expression (Hogg *et al.*, 1987a)

$$(\bar{\bar{C}}_A - \bar{C}_A)/(\bar{\bar{C}}_A - \bar{C}_A)_i = 1 + [k_{AI}C_I/(1 + pk_{AX}\bar{\bar{C}}_A)] \qquad (5.22)$$

where the subscript i signifies an experimental measurement in the presence of competing antigen (I) and where p denotes the antibody valence ($p = 2$ for immunoglobulin G). C_I may be approximated by $\bar{C}_I$, the total antigen concentration, if $\bar{C}_I >> p\bar{C}_A$. Unequivocal evaluation of the intrinsic binding constant for the competing antibody–antigen interaction (k_{AI}) from the linear dependence of the ratio of bound concentrations upon the concentration of competing antigen clearly requires knowledge of k_{AX}. This parameter may be obtained by analyzing measurements of antibody partitioning in the absence of antigen in terms of a Scatchard plot, the slope of which defines the product pk_{AX} (Nichol *et al.*, 1974b; Hogg and Winzor, 1987b; Hogg *et al.*, 1987a).

Only a general guide can be given to the conduct of an experiment for evaluation of the binding constant for an antigen–antibody interaction by ELISA, because the appropriate ranges of antibody and antigen concentrations to be used clearly depend upon the strength of the interaction being studied. Basically, a conjugate of antigen and a protein is first prepared by standard covalent coupling chemistry. Immobilized antigen sites are then created by nonspecific adsorption of the protein moiety of the conjugate to the surface of ELISA plate wells. After removal of excess conjugate by repeated washing steps, reaction mixtures containing a fixed concentration of antibody and a range of antigen concentrations are placed in half of the coated wells. To the remainder is added a range of antibody concentrations to provide the data for evaluating k_{AX}. The ELISA plate is then allowed to incubate for a sufficient time period to attain chemical equilibrium, after which all antigen and antibody in the liquid phase are removed by copious washing steps. The amount of bound antibody in each ELISA well is then monitored by complex formation with a conjugate of anti-immunoglobulin and an enzyme (usually either horseradish peroxidase or alkaline phosphatase) to allow quantification ($\bar{\bar{C}}_A - \bar{C}_A$) on the basis of the enzyme-catalyzed breakdown of a saturating concentration of chromogenic substrate (*o*-phenylenediamine or *p*-nitrophenylphosphate).

Application of this procedure to ELISA data on the interaction between

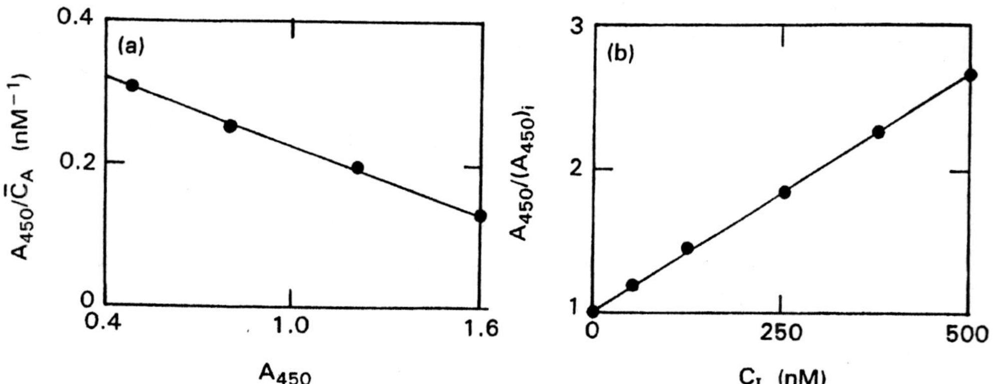

Fig. 5.7. Thermodynamic characterization of an antibody–antigen interaction by quantitiative analysis of an ELISA for paraquat. **(a)** Evaluation of k_{AX} from results for the monoclonal antibody (A) in the absence of competing antigen. **(b)** Plot of results obtained with 24 nM antibody and a range of paraquat concentrations, C_I, in accordance with eq. 5.16. [Adapted from Johnston *et al.* (1988).]

paraquat and a mouse monoclonal antibody (IgG) elicited in response to this univalent antigen (Johnston *et al.*, 1988) is summarized in Figure 5.7, where the concentrations of bound antibody are recorded in terms of absorbances reflecting catalysis by horseradish peroxidase conjugated to the antimouse immunoglobulin. A value of 2.1 ($\pm$0.2) $\times$ 10^8 M^{-1} for interaction of the bivalent monoclonal antibody (IgG) with immobilized paraquat sites (X) is obtained from the slope ($-2k_{AX}$) of the Scatchard plot of results in the absence of competing antigen (Fig. 5.7a). Its combination with the slope $[k_{AI}/(1 + 2k_{AX}\overline{\overline{C}}_A)]$ of the plot for antibody–antigen mixtures (24 nM IgG and 0.05–10 μM paraquat) in accordance with eq. 5.22 (Fig. 5.7b) yields an intrinsic binding constant (k_{AI}) of 2.3 ($\pm$0.2) $\times$ 10^7 M^{-1} for the interaction between paraquat and its elicited monoclonal antibody (Johnston *et al.*, 1988).

A requirement of this affinity chromatographic approach to the characterization of an antibody–antigen interaction by solid-phase immunoassay is that the antigen (I) be univalent. However, essentially the same procedure also applies to studies involving a multivalent antigen provided that the univalent Fab fragment is substituted for the monoclonal antibody (Hogg *et al.*, 1987a). The use of solid-phase immunoassays for rigorous thermodynamic characterization of antibody–antigen interactions is still in its infancy; but future applications have the potential to add an extra dimension to the utility and scope of this widely used immunochemical technique.

6

SIGMOIDAL BINDING CURVES: ANALYSIS AND INTERPRETATION

In the preceding four chapters on experimental methods for quantitative measurement of ligand binding, the emphasis has been on experimental systems for which binding is rectangular hyperbolic. Indeed, apart from the brief consideration of acceptors with more than one class of binding site (Section 1.4 and Fig. 1.2), the focus has been on ligand binding that can be described by a single intrinsic association constant. Attention is now turned toward systems for which the binding of a ligand is not describable in such terms. The major considerations concern the effects of ligand binding to an acceptor that can exist in more than one state, each with a characteristic affinity for ligand. The most striking consequence of this interplay between ligand binding and equilibrium coexistence of the acceptor in multiple ligand-bound states is the sigmoidal rather than rectangular hyperbolic dependence of the binding function (or consequent response) upon ligand concentration that plays such an important role in the control of metabolism by allosteric regulation. However, before embarking on that endeavor it is helpful to consider a simpler system in order to introduce the manner in which interplays of equilibria are to be treated.

6.1. INTERPLAYS OF EQUILIBRIA: LINKED FUNCTIONS

We consider initially the effect of a second ligand (effector, E) on the binding of ligand of prime interest (S) to acceptor (A). This problem of analyzing results in situations where two different ligands bind to an acceptor is frequently addressed by resorting to linked-function theory (Wyman, 1964; Weber, 1975, 1992), a textbook account of which has been presented by Cantor and Schimmel (1980). Its basis is most readily seen by noting that the general equation for the binding of a single ligand to p sites on an acceptor (eq. 1.8) may be written as

$$r = d(\ln \Xi)/d(\ln C_S) \tag{6.1a}$$

$$\Xi = 1 + \sum_{i=1}^{p} (\mathbf{K}_i C_S^i) \tag{6.1b}$$

In these expressions Ξ, the denominator of eq. 1.8, is referred to as the grand partition function. This ratio of total to free acceptor concentrations is a polynomial in free ligand concentration, with the successive coefficients ($\mathbf{K}_i = \Pi K_i$) identified as stoichiometric binding constants for the formation of AS_i from the two reactants A and S.

Introduction of an effector ligand (E) is incorporated by writing the equilibrium constant for the reaction

$$A + iS + jE \rightleftarrows AS_iE_j \tag{6.2a}$$

as

$$\mathbf{K}_{ij} = C_{AS_i}E_j/(C_A C_S^i C_E^j) \tag{6.2b}$$

whereupon the binding functions for ligand (r_S) and effector (r_E) become

$$r_S = \sum_{i=1}^{p} \sum_{j=1}^{q} (iC_{AS_i}E_j) \Big/ \sum_{i=1}^{p} \sum_{j=1}^{q} (C_{AS_i}E_j) \tag{6.3a}$$

$$r_E = \sum_{i=1}^{p} \sum_{j=1}^{q} (jC_{AS_i}E_j) \Big/ \sum_{i=1}^{p} \sum_{j=1}^{q} (C_{AS_i}E_j) \tag{6.3b}$$

for an acceptor with p and q sites for S and E, respectively. By analogy with eqs. 1.8 and 6.1, these equations for the binding function may also be written as

$$r_S = [\partial(\ln \Xi)/\partial(\ln C_S)]_{C_E} \tag{6.4a}$$

$$r_E = [\partial(\ln \Xi)/\partial(\ln C_E)]_{C_S} \tag{6.4b}$$

$$\Xi = \sum_{i=0}^{p} \sum_{j=0}^{q} \mathbf{K}_{ij} C_S^i C_E^j \quad ; \quad \mathbf{K}_{00} = 1 \tag{6.4c}$$

The term *linked function theory* originates from the fact that these expressions give rise to the interrelationships

$$[\partial r_S/\partial(\ln C_E)]_{C_S} = [\partial r_E/\partial(\ln C_S)]_{C_E} \tag{6.5a}$$

$$[\partial(\ln C_E)/\partial(\ln C_S)]_{r_E} = -(\partial r_S/\partial r_E)_{C_S} \tag{6.5b}$$

Thus, at constant concentration of free ligand (C_S), an increase in C_E changes the amount of bound S by an amount exactly equal to the change in amount of bound effector accompanying an increase in ligand concentration at fixed concentration of free effector.

To this point the analysis has been model-independent. However, that thermodynamic rigor is usually forfeited in any attempted mechanistic interpretation of results. The problem arises from an inability to obtain an adequate definition of the grand partition function from the dependence of r_S upon ln C_S at fixed C_E because of the unavoidable uncertainty inherent in estimates of a large and unknown number of stoichiometric binding constants ($\mathbf{K}_{ij}$) obtained by polynomial curve-fitting. This difficulty is therefore obviated by specifying the species (AS_iE_j) that contribute to the grand partition function. By identifying and restricting the number of $\mathbf{K}_{ij}$ for which values are required, such action certainly achieves the aim of decreasing the uncertainty inherent in their determined magnitudes; but it also renders the analysis model-dependent.

Although the use of linked-function theory is undeniably an extremely elegant means of treating interplays of equilibria, familiarity with calculus is a prerequisite for complete understanding of the method. Because resort has to be made to analysis in terms of a specific model anyway, that calculus requirement can be avoided by the following approach, which therefore seems more appropriate for what is presumed to be a largely biological readership. It is introduced by considering the effect of ATP on the binding of proflavin to a single site on thrombin, a system for which the experimental results (De Cristofaro *et al.*, 1990) have already been subjected to analysis by a variant of linked-function theory (Weber, 1975, 1992) in which the stoichiometric binding constants of the grand partition function (Ξ) are expressed in terms of standard free energy changes, exp $(-\Delta\mathbf{G}^0_{ij}/\mathbf{RT})$, rather than $\mathbf{K}_{ij}$.

From the form of the dependence of the binding function for proflavin (r_S) upon ATP concentration (Fig. 6.1), the simplest model that can account for such activation and inhibition by ATP (effector, E) is one in which the enzyme possesses one site for ligand binding but two sites for the effector: the ligand-binding site at which the interactions of S and E are competitive; and an effector site, the occupancy of which by ATP enhances the binding of proflavin to the ligand-binding site. The mechanism for such a model, also invoked by De Cristofaro *et al.* (1990), can be represented as

$$\begin{array}{ccc}
AE \overset{k_{AE}}{\rightleftharpoons} & E + A + S & \overset{k_{AS}}{\rightleftharpoons} AS \\
 & + & \\
 & E & \\
 & \updownarrow k_{EA} & \\
EAE \overset{k_{EAE}}{\rightleftharpoons} & E + EA + S & \overset{k_{EAS}}{\rightleftharpoons} EAS
\end{array} \tag{6.6}$$

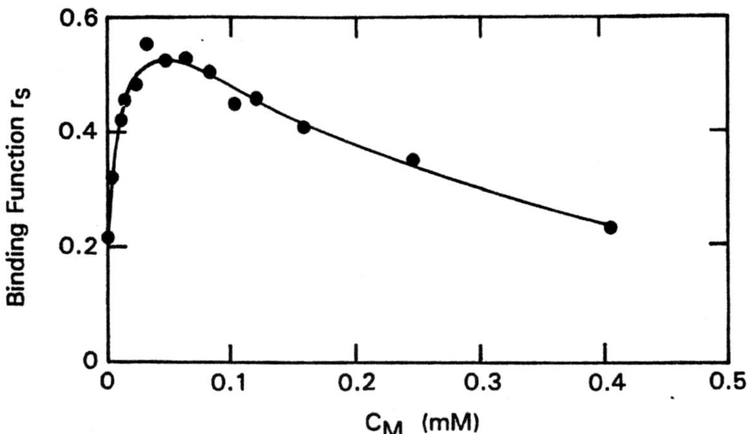

Fig. 6.1. Effect of ATP concentration on the interaction of proflavin with thrombin. [Adapted from De Cristofaro *et al.* (1990).]

where the competitive interactions of proflavin (S) and ATP (E) with the proflavin-binding site are described by k_{AS} and k_{AE}, respectively. Interaction of ATP with the effector site of the enzyme gives rise to an enzyme–effector complex EA, which in turn binds proflavin to form the ternary complex EAS. The respective binding constants are k_{EA} and k_{EAS}. Finally, formation of the ternary complex with ATP attached to both enzyme sites is governed by binding constant k_{EAE}.

Application of the law of mass action to each equilibrium reaction leads to the following expression for the binding function for ligand (r_S):

$$r_S = \frac{C_{AS} + C_{EAS}}{C_A + C_{AS} + C_{EAS} + C_{AE} + C_{EA} + C_{EAE}} \tag{6.7a}$$

$$= \frac{k_{AS}C_S + k_{EA}k_{EAS}C_E C_S}{1 + k_{AS}C_S + (k_{AE} + k_{EA})C_E + k_{EA}k_{EAS}C_E C_S + k_{EA}k_{EAE}C_E^2} \tag{6.7b}$$

Although this equation for the ligand binding function is expressed in terms of five equilibrium constants, there are only four independent polynomial coefficients—a feature evident from reversion to formation constants $\mathbf{K}_{ij}$ (see eq. 6.8).

$$r_S = \frac{\mathbf{K}_{10}C_S + \mathbf{K}_{11}C_E C_S}{1 + \mathbf{K}_{10}C_S + \mathbf{K}_{01}C_E + \mathbf{K}_{11}C_E C_S + \mathbf{K}_{02}C_E^2} \tag{6.8}$$

At this stage the present approach and the linked-function treatment have in fact merged, a factor evident from comparison of the denominator of eq. 6.8

(which is the grand partition function [Ξ] of linked-function theory) with eq. 6 of De Cristofaro *et al.* (1990).

On the basis of the best-fit description of the experimental results in terms of eq. 6.8 (the curve in Fig. 6.1), De Cristofaro *et al.* (1990) have assigned the following magnitudes to the various formation constants: $\mathbf{K}_{10} = k_{AS} =$ 6,200 M^{-1}; $\mathbf{K}_{11} = k_{EA}k_{EAS} = 1.19 \times 10^9$ M^{-2}; $\mathbf{K}_{01} = k_{AE} + k_{EA} =$ 13,800 M^{-1}; $\mathbf{K}_{02} = k_{EA}k_{EAE} = 2.88 \times 10^8$ M^{-2}. Apart from unequivocal evaluation of the binding constant for the enzyme–proflavin interaction (k_{AS}), the only other information available on individual binding constants is that $k_{EAS}/k_{EAE} = \mathbf{K}_{11}/\mathbf{K}_{02} = 4.1$.

Any attempt to delineate the change in enzyme affinity for proflavin as the result of ATP attachment to the effector site must entail the introduction of an assumption about the relative magnitudes of some of the k_i. For example, consideration that only the binding constant for proflavin attachment undergoes change in response to ATP attachment at the effector site allows k_{EA} and k_{EAE} to be equated. However, combination of the experimental value of $\mathbf{K}_{01} = k_{AE} + k_{EA}$ with that for $\mathbf{K}_{02} = k_{EA}k_{AE}$ leads to the physically unacceptable consequence that either k_{EA} or k_{AE} is negative. That model is therefore precluded. Another possible option is to consider that the same binding constant applies to the interactions of ATP with the proflavin-binding and effector sites ($k_{EA} = k_{AE}$). This course of action leads to a value of 6,900 M^{-1} ($\mathbf{K}_{01}/2$) for k_{EA} and k_{AE}, whereupon $k_{EAS} = \mathbf{K}_{11}/k_{EA} =$ 172,500 M^{-1} and $k_{EAE} = \mathbf{K}_{02}/k_{EA} =$ 41,700 M^{-1}. For that model the interaction of ATP with the effector site gives rise to a 28-fold enhancement of proflavin binding but only a 6-fold enhancement of nucleotide binding to the ligand-binding site on thrombin. On the other hand, a more logical simplification might be to consider that the conformational change undergone by the enzyme in response to binding of ATP at the effector site exerts similar activating effects on the binding of proflavin and effector to the ligand-binding site. Adoption of this assumption leads to the conclusion that $k_{AS}/k_{AE} = k_{EAS}/k_{EAE} = 4.1$, whereupon $k_{AE} =$ 1,550 M^{-1} and $k_{EA} = \mathbf{K}_{01} - k_{AE} =$ 12,250 M^{-1}. From the consequent inference that $k_{EAS} = \mathbf{K}_{11}/k_{EA} =$ 187,000 M^{-1}, it follows that $k_{EAS}/k_{AS} = 30$; *i.e.*, that attachment of ATP to the effector site results in 30-fold enhancement of proflavin binding to thrombin.

These two illustrative examples emphasize the model-dependent nature of conclusions emanating from the attempted thermodynamic characterization of even a relatively simple interplay of equilbria. In that regard the decision to adopt a simpler approach is not the cause of the ambiguity of interpretation, which also extends to the analysis in terms of linked-function theory. The problem arises from the fact that the number of equilibrium constants required to describe the mechanism exceeds the number of independent parameters in the binding equation (or grand partition function) on which the analysis is based. This demonstration of such a dilemma in an extremely simple acceptor system with only two sites, each of which is functionally different, highlights (1) the fact that any analysis of results can only establish consistency with interpretation in terms of a given model and (2) the desirability of seeking information additional to binding data in any attempt to characterize more complicated interplays of equilibria.

6.2. POSITIVE AND NEGATIVE COOPERATIVITY

The above thrombin–proflavin–ATP interplay has provided an example of the situation in which the binding of an effector (ATP) to an effector site induces a change in the conformational state of the acceptor (thrombin), thereby enhancing the binding of ligand (proflavin) to its site on the acceptor. If the acceptor comprises several identical (or very similar) subunits, the binding of ligand to any one subunit can induce conformational changes in adjacent subunits and thereby alter their intrinsic affinities for ligand. In other words, the one ligand, S, can act as both a ligand and an effector. This concept of equivalent but dependent sites was introduced by Koshland *et al.* (1966), who realized the existence of two potential situations: one in which the ligand-induced conformational changes in the acceptor increase the magnitude of the intrinsic binding constants for successive interactions of ligand ($k_{i+1} > k_i$) and the other in which the effect of the conformational changes is to diminish the binding affinity of remaining unoccupied sites ($k_{i+1} < k_i$). The former situation, designated as *positive cooperativity*, gives rise to a sigmoidal binding response (Fig. 6.2a), whereas the latter situation, referred to as *negative cooperativity*, yields a binding response of the same focus as that reflecting heterogeneity of acceptor sites (Fig. 6.2b; *cf.* Fig. 1.2). Unfortunately, it is virtually impossible

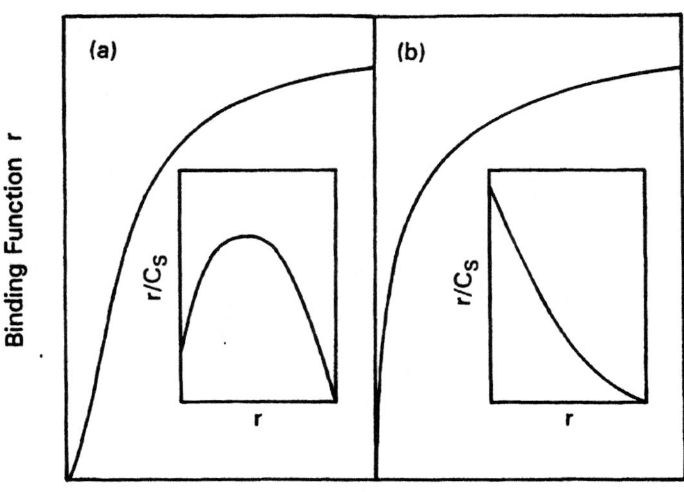

Fig. 6.2. Basic forms of binding curves and their Scatchard transforms (insets) reflecting (a) positive cooperativity and (b) negative cooperativity.

to ascertain whether acceptor sites exhibit heterogeneity in the first place or whether the heterogeneity is ligand induced—a circumstance responsible for the fact that there are no examples of systems in which negative cooperativity of ligand binding has been established unequivocally. At one stage the binding of NAD^+ to glyceraldehyde-3-phosphate dehydrogenase was considered to reflect negative cooperativity (Conway and Koshland, 1968) on the grounds that the four coenzyme-binding sites on the tetrameric enzyme must surely be equivalent. However, subsequent X-ray crystallographic studies have established the existence of coenzyme-site heterogeneity in the apoenzyme (Murthy *et al.*, 1980). We shall therefore concentrate on ligand-induced transitions giving rise to positive cooperativity.

As noted above, a sigmoidal binding response is encountered when the magnitudes of the intrinsic or site-binding constant for the remaining unoccupied sites of acceptor increases with increasing extent of acceptor-site occupancy ($k_{i+1} > k_i$). Allowance for this nonidentity of successive site-binding constants is effected by incorporating the relationships for stoichiometric constants (K_i) in terms of k_i (eq. 1.8) into the general binding expression (eq. 1.7). For a tetrameric acceptor (protein) with four identical subunits ($p = 4$) we thus obtain (Nichol and Winzor, 1981)

$$r = \frac{4k_1C_S + 12k_1k_2C_S^2 + 12k_1k_2k_3C_S^3 + 4k_1k_2k_3k_4C_S^4}{1 + 4k_1C_S + 6k_1k_2C_S^2 + 4k_1k_2k_3C_S^3 + k_1k_2k_3k_4C_S^4} \tag{6.9}$$

In the model of allostery based on the concept of positive cooperativity, Koshland *et al.* (1966) ascribe a more detailed mechanistic meaning to the site-binding constants k_i ($1 \le i \le 4$) by expressing them in terms of four other parameters. k_S is the intrinsic binding constant for interaction of ligand with an individual subunit of ligand-free acceptor in which all subunits are in conformational state A; and Y is the isomerization constant for the conformational transition from a subunit in state A to state B: k_{AB} and k_{BB}, the constants used to define the subunit interactions that give rise to nonequivalence of successive binding ($k_{AA} = 1$), are weighted in the expressions for k_i in accordance with the number of contacts between subunits in the specified conformational states.

This aspect of the Koshland approach becomes evident from inspection of Figure 6.3, which lists the expressions that replace the Πk_i of eq. 6.9 for various geometrical arrangements of the four subunits. The problem at hand in an experimental context is to select the geometry and values of k_{AB}, k_{BB}, and the product Yk_S that best describe the polynomial coefficients Πk_i in the general binding expression (eq. 6.9). More direct procedures for the evaluation of these parameters have been devised (Koshland *et al.*, 1966; Koshland and Neet, 1968). However, although resorting to such procedures may well allow the assignment of magnitudes to parameters for the geometrical arrangement of subunits that best describes the binding curve, such endeavors must be viewed in the light that the various Πk_i in eq. 6.9 may also be amenable to other interpretations.

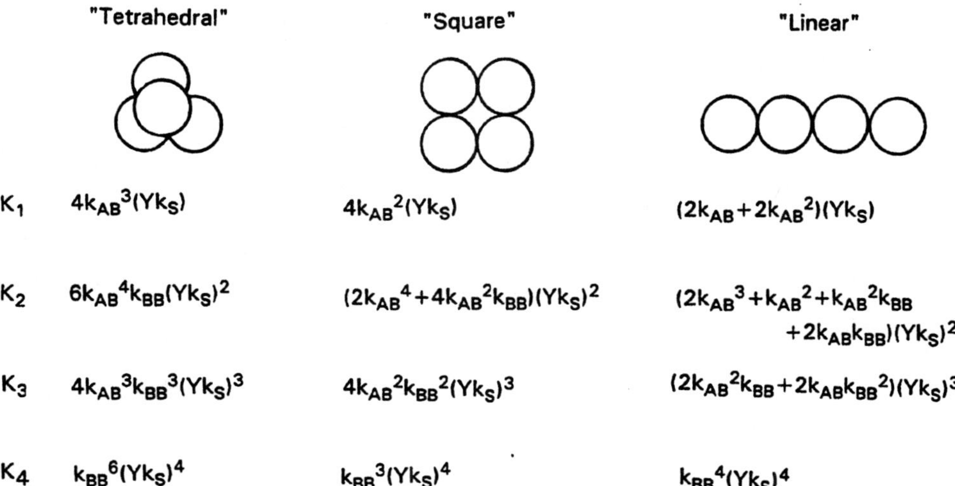

Fig. 6.3. Relationships between the stoichiometric binding constants (K_i) and the Koshland cooperativity parameters for three geometrical arrangements of a tetrameric acceptor with a binding site for ligand on each subunit.

6.3. THE MONOD MODEL OF ALLOSTERY

Instead of considering the acceptor to undergo ligand-induced conformational changes, Monod *et al.* (1965) have proposed that the flexibility of protein structure may well be commensurate with the equilibrium coexistence of two isomeric states of the acceptor in an enzyme solution prior to the introduction of ligand. Although the binding sites within an acceptor state are considered to be equivalent and independent, the intrinsic binding constants and/or numbers of binding sites are allowed to differ between states because of their different protein conformations. For an equilibrium mixture of two isomeric forms, R and T, possessing p and q ligand-binding sites, respectively, governed by intrinsic association constants k_{RS} and k_{TS} (Fig. 6.4a), the binding equation for each isomer assumes the unsimplified form of eq. 1.9a, whereupon it follows that

$$r = \frac{pk_{RS}C_S(1 + k_{RS}C_S)^{p-1} + Xqk_{TS}C_S(1 + k_{TS}C_S)^{q-1}}{(1 + k_{RS}C_S)^p + X(1 + k_{TS}C_S)^q} \tag{6.10}$$

in which $X = C_T/C_R$ is the equilibrium constant describing the pre-existing isomerization of acceptor. The equivalence of this binding equation with that (eq. 6.9) for positively cooperative binding is emphasized by writing eq. 6.10 as a ratio of polynomials for a system with $p = q = 4$. The result is

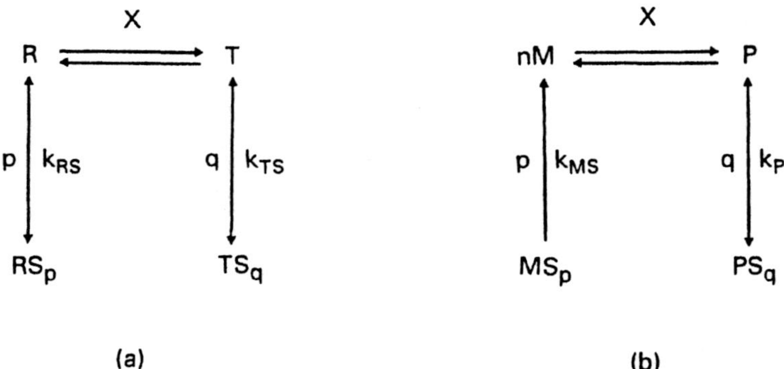

Fig. 6.4. Allosteric models based on the preferential binding of ligand to acceptors undergoing **(a)** pre-existing isomerization (Monod *et al.*, 1965) and **(b)** pre-existing self-association (Nichol *et al.*, 1967a).

$$r = \frac{4(k_{RS} + Xk_{TS})C_S + 12(k_{RS}^2 + Xk_{TS}^2)C_S^2 + 12(k_{RS}^3 + Xk_{TS}^3)C_S^3 + 4(k_{RS}^4 + Xk_{TS}^4)C_S^4}{(1 + X) + 4(k_{RS} + Xk_{TS})C_S + 6(k_{RS}^2 + Xk_{TS}^2)C_S^2 + 4(k_{RS}^3 + Xk_{TS}^3)C_S^3 + (k_{RS}^4 + Xk_{TS}^4)C_S^4} \tag{6.11}$$

Comparison of eqs. 6.9 and 6.11 reveals these identities: $k_1 \equiv (k_{RS} + Xk_{TS})/(1 + X)$; $k_1k_2 \equiv (k_{RS}^2 + Xk_{TS}^2)/(1 + X)$; $k_1k_2k_3 \equiv (k_{RS}^3 + k_{TS}^3)/(1 + X)$; $k_1k_2k_3k_4 \equiv (k_{RS}^4 + k_{TS}^4)/(1 + X)$. Consequently, binding curves obtained for an acceptor undergoing pre-existing isomerization will also be sigmoidal (as in Fig. 6.2a), except for the special situation in which the isomeric states exhibit identical numbers of sites and affinities for ligand ($p = q$ and $k_{RS} = k_{TS}$). On substitution of those identities, eq. 6.10 simplifies to a rectangular hyperbolic relationship.

The formal identity of eqs. 6.9 and 6.11 means that any characterization of a sigmoidal binding (or kinetic) response in terms of either the Koshland or the Monod model of allostery must be equivocal, because there is always an equivalent interpretation in terms of the other concept. A possible way of avoiding this ambiguity of interpretation is to test for the existence of an isomerization equilibrium in the absence of ligand. For rabbit muscle pyruvate kinase the pre-existence of an isomerization equilibrium has been demonstrated by using the molecular crowding effect of an inert solute (sucrose) to displace the equilibrium position toward the smaller isomeric state — a phenomenon detected by sedimentation velocity studies (Harris and Winzor, 1988b). Unfortunately, that approach may not be definitive either, because failure to observe any change in the size parameter being monitored does not preclude the existence of an isomerization equilibrium. It may merely signify that the conformational change is too subtle to be detected as a gross change in size and/or shape. Furthermore, the two models are not necessarily mutually exclusive inasmuch as coexisting acceptor states may also exhibit cooperativity of ligand binding (Saroff, 1991).

In view of the above considerations, a realistic experimental approach is to forego aspirations of achieving an unequivocal and mechanistically correct thermodynamic characterization of a sigmoidal binding response and to concentrate on obtaining an acceptable description of the data in terms of the allosteric model for which parameters may be evaluated more easily. Our inclination, albeit possibly biased because of greater familiarity with one model, is to favor an interpretation in terms of the Monod concept, which may also be adapted readily to incorporate the effects of allosteric effectors.

Consider the binding of a ligand and an effector, S and E, to different sites on the R isomer and also to distinct sites on the T isomer. Under these circumstances eq. 6.10 continues to describe the binding of ligand (S) provided that the isomerization constant is replaced by a constitutive isomerization constant, X', where

$$X' = X(1 + k_{TE})^z/(1 + k_{RE})^w \tag{6.12}$$

In this expression w and z denote the respective numbers of acceptor sites for E on R and T isomers, to which intrinsic binding constants k_{RE} and k_{TE} apply (Frieden, 1967; Changeux and Rubin, 1968; Nichol *et al.*, 1972b). Preferential binding of effector (E) to the isomer with lower affinity for S gives rise to allosteric inhibition and an increasingly sigmoidal binding curve as $\bar{C}_E$ is raised, whereas the binding curve will show activation and an increasing tendency to become rectangular hyperbolic if E binds preferentially to the more active isomer. These effects are evident in Figure 6.5, which illustrates the inhibitory effect of CTP and the activating effect of ATP on the kinetics of conversion of aspartate to carbamyl aspartate by aspartate transcarbamylase in the presence of a saturating concentration of carbamyl phosphate. The lines in Figure 6.5 represent analyses of experimental data (Gerhart, 1970) in terms of eqs. 6.10 and 6.12 with $p = q = w = z = 6$ and with v/v_m substituted for r/p (see Section 1.6) for enzyme with the T isomer inactive ($k_{TS} = 0$). Curve-fitting of results in the absence of either effector has led to the conclusion that $k_{RS} = 180\ M^{-1}$ and $X = 8$ (Smith, 1977), whereas the increased sigmoidality of results in the presence of 0.5 mM CTP may be accommodated by assigning respective magnitudes of 5.1×10^4 and $7.4 \times 10^4\ M^{-1}$ to k_{RE} and k_{TE}, thereby increasing the magnitude of X' (eq. 6.12). The less sigmoidal form of the curve in the presence of 2 mM ATP may be rationalized by invoking values of 10^5 and $10^2\ M^{-1}$ for k_{RE} and k_{TE}, respectively, to obtain the required decrease in X' (Smith, 1977).

The isomerizing model has the ability to provide a readily evaluated quantitative characterization of a sigmoidal response in terms of notional parameters for the assigned model (mechanism) used to portray the interplay of equilibria. However, it should be noted that the use of only three interaction parameters

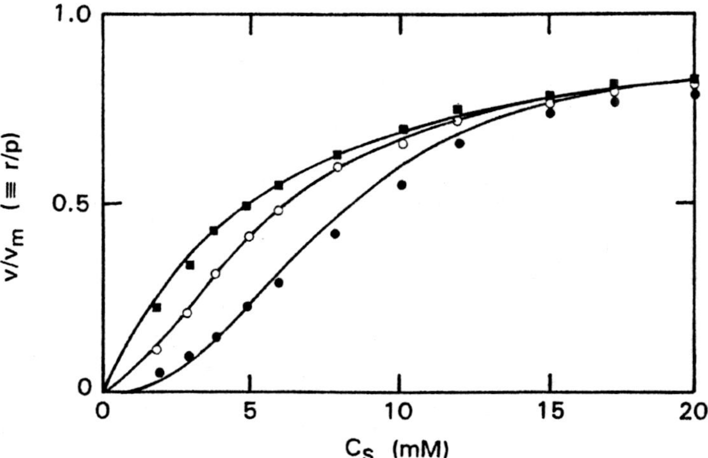

Fig. 6.5. Allosteric activation and inhibition of aspartate transcarbamylase by effectors. ○, Dependence of initial velocity upon aspartate concentration, C_S, in the presence of saturating carbamyl phosphate; ●, ■, corresponding dependences with CTP (0.5 mM) and ATP (2 mM), respectively, also included in the reaction mixtures. [Adapted from Smith (1977).]

(k_{RS}, k_{TS}, X) for a model with $p = q$ (eq. 6.10) does place restrictions on the inter-relationships between the Πk_i coefficients of eq. 6.9: a restriction that can be alleviated to some extent by allowing the number of sites on the two isomeric states to differ ($p \neq q$) or by extending the model to include additional states in the pre-existing isomerization equilibrium (Baghurst *et al.*, 1978; Nichol and Winzor, 1981). The point to bear in mind is that the aim of the quantitative characterization of ligand binding is primarily to establish the adequacy of a particular model as a thermodynamic description of the binding behavior. In that regard the fact that a model may be incorrect mechanistically does not detract from use of the evaluated parameters to predict the distribution of ligand between free and acceptor-bound states as a function of total ligand and acceptor concentrations under otherwise identical experimental conditions to those used for the characterization.

Finally, although models involving isomerization of acceptor (be it ligand induced or pre-existing) provide considerable flexibility from the viewpoint of describing a sigmoidal dependence of binding response upon free ligand concentration, it should be noted that the binding expressions (eqs. 6.9 and 6.10) contain no term in acceptor concentration and hence that these models cannot provide an adequate thermodynamic description of systems in which the binding function at a particular C_S is a function of acceptor concentration. We therefore explore an extension of the Monod model in which the pre-existing equilibrium entails self-association rather than isomerization of acceptor.

6.4. ALLOSTERIC MODELS BASED ON ACCEPTOR SELF-ASSOCIATION

Because many proteins comprise an equilibrium mixture of monomeric and higher polymeric states, the Monod model has been adapted to include the preferential binding of ligand to an acceptor undergoing pre-existing self-association (Nichol *et al.*, 1967a). For a two-state self-associating acceptor ($nM \rightleftarrows P$) with p and q sites on monomer and polymer, respectively (Fig. 6.4b), the binding equation is

$$r = \frac{pk_{MS}C_S(1 + k_{MS}C_S)^{p-1} + XC_M^{n-1}qk_{PS}C_S(1 + k_{PS}C_S)^{q-1}}{(1 + k_{MS}C_S)^p + nXC_M^{n-1}(1 + k_{PS}C_S)^{q-1}} \qquad (6.13)$$

where k_{MS} and k_{PS} are the respective intrinsic binding constants for the interactions of ligand with monomer and polymer, whose free concentrations are governed by an association equilibrium constant $X = C_P/C_M^n$. The first point to note is that when $n = 1$ (acceptor isomerization), eq. 6.13 becomes eq. 6.10, which emphasizes that the Monod model may be regarded as a special case of a more general model based on pre-existing acceptor self-association. Indeed, it is a very special case inasmuch as selection of a value of unity for n eliminates the terms in C_M. If the pre-existing equilibrium entails self-association in ($n > 1$), then the binding equation and also the grand partition function (Ξ) can no longer be written in terms of C_S as the single variable. Consequently, whereas the binding function has thus far been regarded as an experimental parameter whose magnitude is governed solely by the free ligand concentration, that frequently presumed generalization does not extend to acceptors undergoing self-association. Instead, binding curves become dependent upon total acceptor concentration, $\overline{C}_A$, each curve at a fixed acceptor concentration being defined by eq. 6.13 with the concentration of free monomer, C_M, given by the physically acceptable solution ($0 < C_M \leq \overline{C}_A$) of the simultaneous equation

$$nX(1 + k_{PS}C_S)^qC_M^n + (1 + k_{MS}C_S)^pC_M - \overline{C}_A = 0 \qquad (6.14)$$

In eq. 6.14, which is the statement of mass conservation, $\overline{C}_A$ is the total base-molar concentration of acceptor, *i.e.*, the weight concentration (g/liter) divided by monomer molecular weight M_M. It is also the concentration of acceptor to be used in experimental determination of the binding function (eq. 1.3).

Figure 6.6 provides three examples of systems for which the binding curves are dependent upon acceptor concentration, each binding curve having been determined with the indicated $\overline{C}_A$. The importance of so doing cannot be overemphasized, because variation of both $\overline{C}_A$ and $\overline{C}_S$ would lead to an experimental binding curve that is a composite of segments from those described by eq. 6.13 for each concentration of acceptor ($\overline{C}_A$) used. Results for the binding

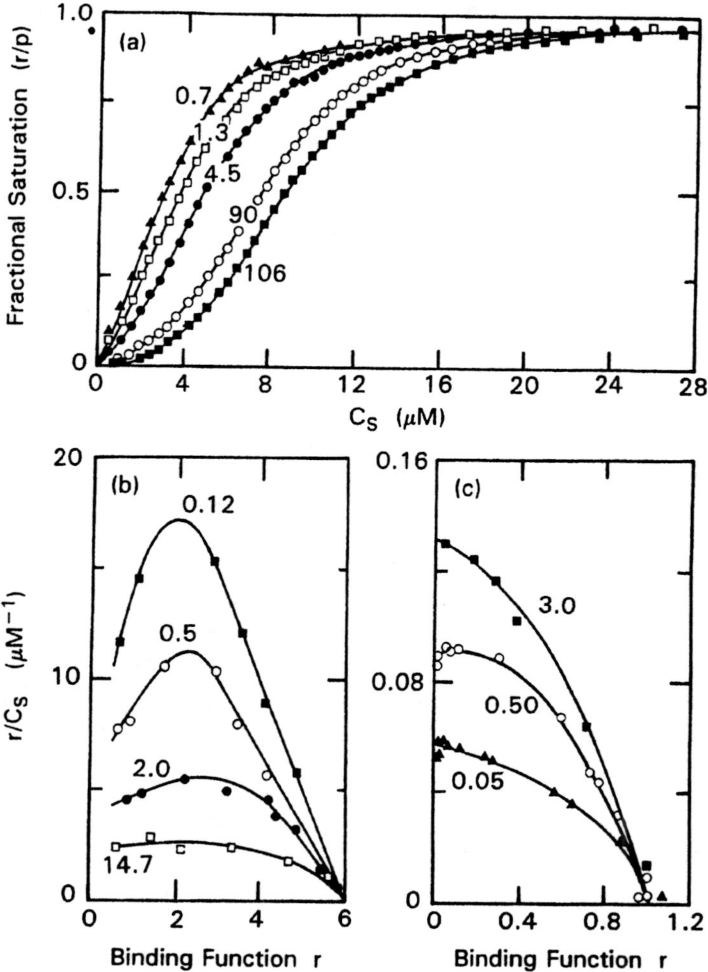

Fig. 6.6. Experimental systems exhibiting dependence of the binding function upon acceptor concentration. **(a)** Binding curves for oxygen uptake by the indicated concentrations (mg/liter) of human hemoglobin (pH 7.4). [Adapted from Mills *et al.* (1976).] **(b)** Scatchard plots for the interaction of GTP with the indicated concentrations (g/liter) of bovine liter glutamate dehydrogenase in the presence of 400 μM NADH (pH 7.1). [Adapted from Frieden and Colman (1967).] **(c)** Corresponding plots for the binding of oxytocin to bovine neurophysin (pH 5.6) at the indicated concentrations (g/liter) of the acceptor. [Adapted from Nicolas *et al.* (1976).]

of oxygen to hemoglobin (Mills *et al.*, 1976) at concentrations where the $\alpha\beta$ and $\alpha_2\beta_2$ species coexist in equilibrium are presented in Figure 6.6a, from which it can be concluded that the $\alpha\beta$ species binds oxygen preferentially. The binding of GTP to glutamate dehydrogenase (Fig. 6.6b) provides another example of a system in which binding is enhanced by a decrease in $\overline{C}_A$, thereby signifying that monomer is the enzyme form that binds ligand preferentially

(Frieden and Colman, 1967). On the other hand, the binding of oxytocin to bovine neurophysin is favored by an increase in acceptor concentration (Fig. 6.6c) and thus exemplifies the situation in which the polymeric state binds ligand preferentially (Nicolas *et al.*, 1976, 1978; Tellam and Winzor, 1980).

For several years it was considered that preferential binding of ligand to an acceptor undergoing reversible self-interaction (isomerization or self-association) could only give rise to binding curves symptomatic of positive cooperativity. Although that conclusion remains true for isomerizing systems, the binding of Zn^{2+} to α-amylase (Fig. 6.7a) demonstrates that curves of the negatively cooperative form can also arise from preferential binding to a self-associating acceptor. For this system the zinc ion is binding exclusively to a single site on dimeric α-amylase (Tellam *et al.*, 1978).

Those results for the binding of Zn^{2+} to α-amylase prompted closer inspection of the requirements for the observation of sigmoidal binding curves. For the simplest situations involving exclusive binding to the monomer ($q = 0$) or polymer ($p = 0$) state, the requirements for existence of a maximum in the Scatchard plot are (Baghurst *et al.*, 1978; Nichol *et al.*, 1979)

$$X\overline{C}_A^{n-1} > \frac{[n^2(p-1)+n]^{n-1}}{[n^2(p-1)]^n} \quad ; \quad q = 0 \qquad (6.15a)$$

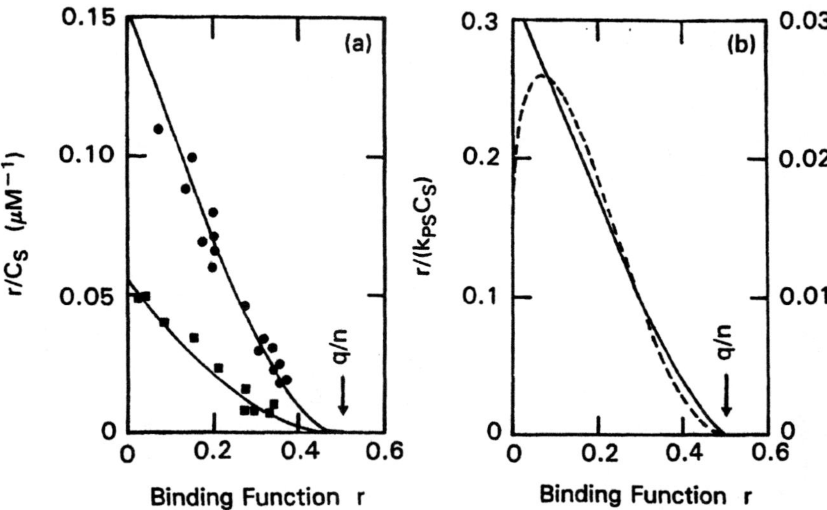

Fig. 6.7. Effects of preferential binding to the polymeric state of an acceptor undergoing reversible self-association in instances where $q < np$. **(a)** Scatchard plots of binding data for the interaction of Zn^{2+} with 1.41 (■) and 6.32 (●) g/liter α-amylase. [Adapted from Tellam *et al.* (1978).] **(b)** Simulated Scatchard plots for exclusive binding of ligand to two sites on tetramer ($q = 2$) of a monomer–tetramer acceptor system with $X = 10^{12}\ M^{-3}$: solid line, $\overline{C}_A = 200\ \mu M$ (left-hand ordinate); dashed line, $\overline{C}_A = 20\ \mu M$ (right-hand ordinate). [Adapted from Nichol *et al.* (1979).]

$$X\overline{C}_A^{n-1} < \frac{(q-1)(n+q-1)^{n-1}}{n^{n+1}} \quad ; \quad p = 0 \qquad (6.15b)$$

For an isomerizing acceptor ($n = 1$) these expressions simplify to $X > 1/(p - 1)$ if $q = 0$ and $X < 1/q$ if $p = 0$. For a self-associating acceptor, however, the condition for a sigmoidal binding response is a function of the dimensionless product $X\overline{C}_A^{n-1}$. An interesting situation arises for systems with $p = 0$ and $1 < q < n$. For a dimerizing system with $q = 1$, the situation encountered in Figure 6.7a, a sigmoidal binding curve can never be obtained because the requirement specified by eq. 6.15b is that $X\overline{C}_A$ be negative. On the other hand, calculations based on eqs. 6.13 and 6.14 have indicated very striking binding behavior for a monomer–tetramer system with two sites on tetramer (Fig. 6.7b). At the lower acceptor concentration the theoretical binding curve is similar to that observed for the binding of Zn^{2+} to α-amylase; but a 10-fold decrease in $\overline{C}_A$ leads to sigmoidal binding behavior. It remains to reiterate the importance of using a fixed acceptor concentration ($\overline{C}_A$) in the experimental determination of a binding curve for a self-associating acceptor, or, indeed, for any acceptor whose self-association characteristics have not been tested.

To conclude this section, brief mention is made of the alternative possibility that acceptor self-association does not occur in the absence of ligand. As in the isomerization case, consideration of the acceptor self-association to be ligand induced rather than pre-existing leads to qualitatively similar binding behavior (Nichol and Winzor, 1976). However, unequivocal distinction between the two possible situations should emanate from the effect of ligand concentration on a weight-average size characteristic (molecular weight, elution volume, or sedimentation coefficient) of the system. Indeed, molecular weight studies have the potential to be far more rewarding than binding studies from the viewpoint of characterizing definitively the interplay between ligand binding and acceptor self-association.

6.5. PERTURBATION OF ACCEPTOR SELF-ASSOCIATION BY LIGAND BINDING

In this section on the effect of a ligand on the macromolecular state of an acceptor undergoing reversible self-association, we shall employ molecular weight as the size parameter being monitored by either sedimentation equilibrium or light-scattering measurements. Although techniques such as frontal gel chromatography (Winzor and Scheraga, 1964) and sedimentation velocity may also be used to detect displacements of self-association equilibria by a ligand, the shifts in the equilibrium position are not as readily quantified because of the lack of a definitive value for the parameter (elution volume or sedimentation coefficient) pertaining to the polymeric acceptor state. Its molecular weight is unequivocally nM_M, the product of the association stoichiometry and the monomer molecular weight.

Provided that an acceptor undergoes pre-existing self-association between monomer and a single polymeric state, a value of the association equilibrium constant, X, may be determined from investigations of acceptor in the absence of the ligand. Because the weight-average molecular weight ($\overline{M}_w$) of an equilibrium mixture with total weight concentration (g/liter) of acceptor, $\bar{c}_A$, is given by

$$\overline{M}_w = [c_M M_M + nM_M(\bar{c}_A - c_M)]/\bar{c}_A \qquad (6.16)$$

where c_i ($i = M$ or P) denotes the weight concentration of species i, the concentration of monomer, c_M, may be determined as the only parameter of unknown magnitude for any assigned value of n. Specifically, $c_m = \bar{c}_A(nM_M - \overline{M}_w)/[(n - 1)M_M]$. Thus, measurement of c_M as a function of total acceptor concentration ($\bar{c}_A$) allows the association constant (X) to be determined as

$$X = [(\bar{c}_A - c_M)/c_M^n][(M_M^{n-1})/n] \qquad (6.17a)$$

This endeavor is frequently effected by writing eq. 6.17a in logarithmic format as

$$\log(\bar{c}_A - c_M) = \log(nX/M_M^{n-1}) + n \log c_M \qquad (6.17b)$$

X may thus be obtained from the ordinate intercept, $\log(nX/M_M^{n-1})$, of the linear plot of $\log(\bar{c}_A - c_M)$ *versus* $\log c_M$, which has a mandatory slope of n. Correctness of the value assigned to n is thereby recognized retrospectively.

The displacement of the association equilibrium may also be examined in similar manner if S is small in comparison with acceptor, whereupon $M_{AS_i} \approx M_M$ and $M_{PS_i} \approx M_P$. In those circumstances eq. 6.16 defines $\bar{c}_M$, the combined weight concentrations of monomer in free and ligand-bound states, and $(\bar{c}_A - \bar{c}_M)$, the corresponding quantity for polymeric acceptor. A constitutive association equilibrium constant, X^* (Baghurst and Nichol, 1975), may therefore be obtained from the expression

$$X^* = (\bar{c}_A - \bar{c}_M)/\bar{c}_M^n = X(1 + k_{PS}C_S)^q/(1 + k_{MS}C_S)^{np} \qquad (6.18)$$

Consequently, provided that exhaustive dialysis of acceptor solution against multiple changes of ligand solution is used to establish the concentration of free ligand (C_S) in the mixtures subjected to analysis, the measurement of X^* as a function of C_S yields information on the relative affinities of the monomer and polymer states of acceptor for ligand.

This aspect of the characterization of ligand binding to a self-associating acceptor is illustrated by a study of the interactions of *N*-acetyltryptophan with α-chymotrypsin, which undergoes reversible dimerization in acetate-chloride buffer, pH 3.9, **I** 0.2 (Winzor and Scheraga, 1964; Aune and Timasheff, 1971;

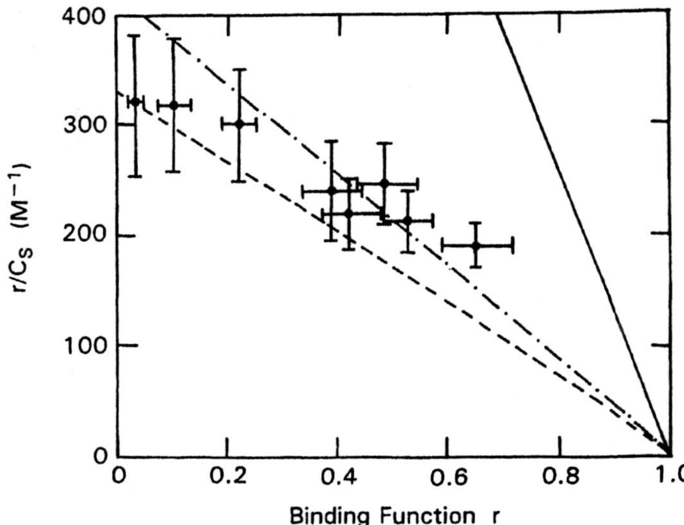

Fig. 6.8. Scatchard plot of binding data for the interaction of *N*-acetyltryptophan with α-chymotrypsin (pH 3.9, **I** 0.2), obtained by frontal gel chromatography with an enzyme concentration of 18.5 g/liter). The broken lines represent theoretical plots based on eqs. 6.13 and 6.14, with $n = 2$, $p = 1$, $q = 2$, $X = 44{,}000\ M^{-1}$, $k_{AS} = 1{,}300\ M^{-1}$ (solid line inferred from kinetic studies), and values of 200 M^{-1} (dashed line) and 300 M^{-1} (dashed, dotted line) for k_{PS}. [Adapted from Tellam *et al.* (1979).]

Horbett and Teller, 1973). The Scatchard plot of data from a binding study with 18.5 mg/ml enzyme (Tellam *et al.*, 1979) is presented in Figure 6.8, together with the corresponding relationship for monomeric enzyme deduced from competitive inhibition of the hydrolysis of *N*-acetyltryptophan ethyl ester by 0.09 mg/ml α-chymotrypsin. The first point to note is that the increase in α-chymotrypsin concentration from an enzymatic level to that used in the binding study has diminished considerably the affinity of the enzyme for the ligand. In keeping with the consequent conclusion that monomeric α-chymotrypsin binds *N*-acetyltryptophan preferentially, the Scatchard plot of the binding data is of the form associated with a sigmoidal response (Fig. 6.2a). Because (1) the magnitudes of p and q are known to be unity and two, respectively (Nichol *et al.*, 1972a), and (2) the value of k_{MS} is available from the competitive inhibition of enzyme catalysis, one approach to the determination of k_{PS} might appear to be nonlinear regression analysis of the binding data on the basis that $X = 44{,}000\ M^{-1}$ (Tellam *et al.*, 1979). However, of the two remaining parameters whose magnitudes need to be determined in eq. 6.13 (k_{PS} and C_M), only k_{PS} is a constant. Consequently, the simplest approach is to assign values to k_{PS}, whereupon eq. 6.14 may be used to calculate c_M (and hence C_M) for each C_S and thereby allow the construction of a simulated binding curve (via eq. 6.13) for each input value of k_{PS}. Adoption of this approach (Tellam *et al.*,

1979) leads to the conclusion that k_{PS} is between 200 and 300 M^{-1} (broken lines in Fig. 6.8).

A more direct approach is to obtain k_{PS} from an analysis of sedimentation equilibrium distributions for enzyme–ligand mixtures obtained by dialyzing α-chymotrypsin (1–3 mg/ml) against many changes of *N*-acetyltryptophan (1–10 mM) in acetate-chloride buffer (pH 3.9, **I** 0.20) to establish the magnitude of C_S appropriate to eq. 6.18. Results of such experiments, plotted in accordance with eq. 6.17b, are summarized in Figure 6.9, about which the following points are noted.

1. Instead of employing the sedimentation equilibrium distribution to determine the weight-average molecular weight for each mixture, the omega function (Milthorpe *et al.*, 1975: see also Section 4.3) has been used to determine the concentration of monomer (and hence of polymer as the difference between total and monomer concentrations) throughout each sedimentation equilibrium distribution.

2. The magnitude of X^* decreases systematically as C_S is varied from 1 mM to 10 mM. These values, together with the corresponding magnitudes of

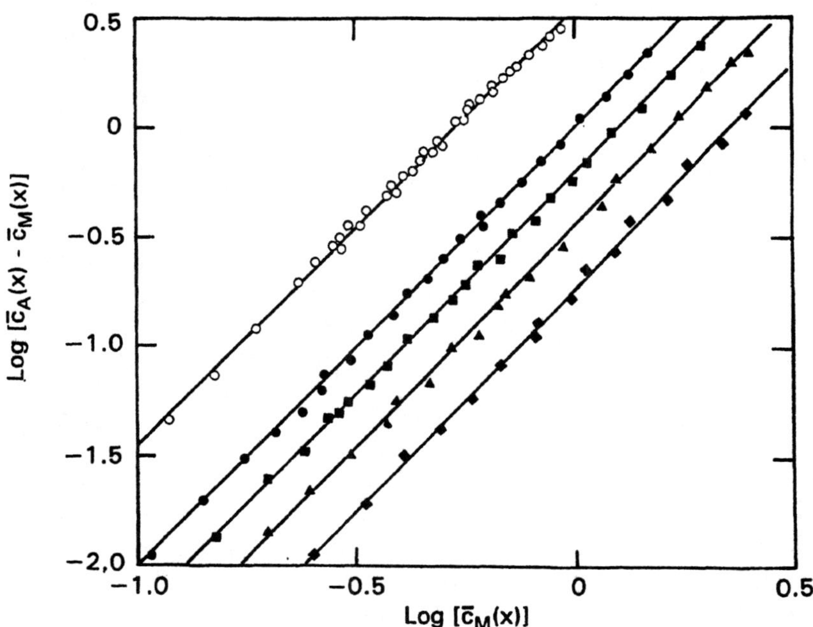

Fig. 6.9. Effect of *N*-acetyltryptophan binding on the dimerization of α-chymotrypsin (pH 3.9, **I** 0.2), the constituent concentration of monomer, $\bar{c}_M(x)$, having been obtained by omega analysis (Milthorpe *et al.*, 1975) of sedimentation equilibrium distributions, $\bar{c}_A(x)$ *versus* x. Results for enzyme dialyzed against buffer (○) and buffer supplemented with 1 mM (●), 2 mM (■), 5 mM (▲), and 10 mM (◆) *N*-acetyltryptophan are plotted in accordance with eq. 6.17b.

k_{PS} deduced from eq. 6.18 with k_{MS} = 1,300 M^{-1} (Fig. 6.8), are as follows: C_S = 10 mM, X^* = 2,300 M^{-1}, k_{PS} = 220 M^{-1}; C_S = 5 mM, X^* = 4,500 M^{-1}, k_{PS} = 280 M^{-1}; C_S = 2 mM, X^* = 7,600 M^{-1}, k_{PS} = 250 M^{-1}; and C_S = 1 mM, X^* = 12,500 M^{-1}, k_{PS} = 230 M^{-1}.

This direct determination of k_{PS} as 250 ($\pm$60) M^{-1} clearly provides a more definitive estimate of the dimer affinity than that inferred from Figure 6.8, which merely signifies a range of 200–300 M^{-1} for k_{PS}.

The determination of k_{PS} via the dependence of X^* upon C_S is not conditional upon access to an analytical ultracentrifuge. Indeed, the binding constant for dimeric α-chymotrypsin was evaluated initially (Tellam *et al.*, 1979) by frontal gel chromatography, which may also be used to define X^*. Specifically, the counterpart of eq. 6.16 is

$$\bar{c}_M = \bar{c}_A(\bar{V}_A - V_p)/(V_M - V_p) \qquad (6.19)$$

where $\bar{V}_A$ is the effluent volume corresponding to the median bisector of the protein constituent in frontal gel chromatography of an enzyme–ligand mixture with predetermined C_S, V_M and V_P being the respective elution volumes of the monomeric and dimeric forms of α-chymotrypsin. The use of a gel chromatographic medium (Bio-Gel P-30) that excluded dimeric α-chymotrypsin allowed the identification of V_P (= V_{PS_i}) as the void volume of the column; and V_M (= V_{MS}) was obtained as the elution volume in the limit of infinite dilution ($\bar{c}_A \to 0$). The availability of these two species elution volumes allows eq. 6.19 to be used to determine the magnitudes of $\bar{c}_M$ required for the application of eq. 6.18 to obtain X^* and hence k_{PS}. Values of 210 and 240 M^{-1} for k_{PS} emanated from frontal gel chromatographic experiments with C_S = 9.17 and 4.77 mM, respectively.

The above example has served to emphasize the advantages of combining binding studies with molecular size measurements to characterize the operative equilibria in an interplay of ligand binding and acceptor self-association. Under some conditions, however, the magnitudes of X, k_{MS}, and k_{PS} may preclude the determination of either k_{MS} or k_{PS} from binding studies. Indeed, that situation would have arisen in the above example except for the fact that advantage could be taken of the enzymatic property of α-chymotrypsin to determine k_{MS} as the inhibition constant in kinetic experiments with a sufficiently low $\bar{c}_A$ for the α-chymotrypsin to be essentially monomeric in the absence of ligand. Fortunately, the perturbing effect of ligand binding on the macromolecular state of the acceptor can still suffice for the delineation of k_{PS} and k_{MS} (albeit with greater uncertainty) in these adverse circumstances. Pertinent to that inference is the recent determination of the binding affinities of the $\alpha\beta$ and $\alpha_2\beta_2$ forms of methemoglobin for NADH solely on the basis of sedimentation equilibrium measurements of the effect of ligand concentration upon X^* (Jacobsen and Winzor, 1995). For this system with $p = q = 1$ and $k_{PS} = 10k_{MS}$, *preferential binding* is an ambiguous term in the sense that the dimer is the preferred acceptor state from the viewpoint of affinity, whereas the monomer is the

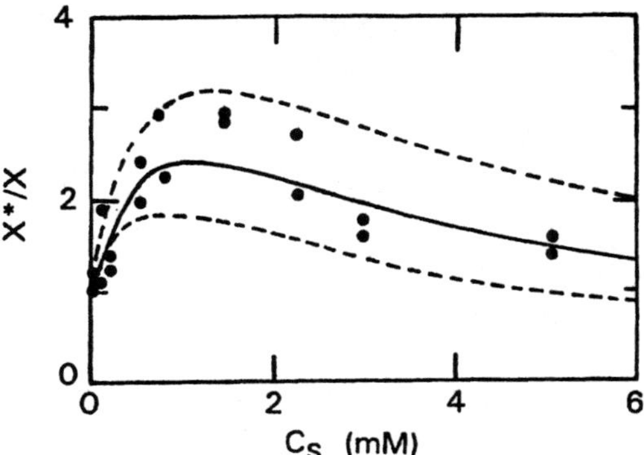

Fig. 6.10. Evaluation of the binding constants for the interactions of NADH with the $\alpha\beta$ and $\alpha_2\beta_2$ states of methemoglobin by sedimentation equilibrium measurements of the constitutive association constant (X^*) as a function of free ligand concentration. The solid line is the best-fit description of the results in terms of eq. 6.16, with $n = 2$, $p = q = 1$; and the broken lines denote the uncertainty envelope associated with the analysis, namely, $k_{MS} = 700\ (\pm 100)\ M^{-1}$ and $k_{PS} = 6{,}000\ (\pm 1{,}000)\ M^{-1}$. [Adapted from Jacobsen and Winzor (1995).]

preferred binding form on the basis of site numbers per base mole of acceptor—a paradox that gives rise to the existence of a maximum in the dependence of X^* upon C_S (Baghurst and Nichol, 1975; Gow *et al.*, 1990). This behavior is evident in the results for the methemoglobin-NADH system (Fig. 6.10).

Studies of the macromolecular state of the acceptor as a function of ligand concentration also have the potential to distinguish systems involving the preferential interaction of ligand with the polymeric state of a self-associating acceptor from those in which an acceptor–ligand complex self-associates. In principle, studies of the concentration dependence of molecular weight of the acceptor in the absence of ligand should suffice to make this distinction. In practice, however, the failure to detect any concentration dependence of molecular weight does not necessarily reflect the absence of self-association ($X = 0$) because it may merely signify that equilibrium coexistence of monomeric and polymeric states of a self-associating acceptor occurs in a range of concentration not covered by the method(s) used to monitor the macromolecular state of the acceptor. There is thus a need for a procedure that allows unequivocal identification of the appropriate model.

For illustrative purposes, the dependence of X^* upon ligand concentration for a system involving preferential binding to the dimeric form of a reversibly associating acceptor is compared with the corresponding dependence for a system in which the dimerization is ligand induced. For the latter situation the appropriate equilibria are

$$M + S \underset{Y}{\overset{k_{MS}}{\rightleftharpoons}} MS \tag{6.20a}$$

$$2MS \rightleftharpoons PS_2 \tag{6.20b}$$

which clearly achieve the same reaction stoichiometry as the preferential binding scheme (Fig. 6.4) with $p = 1$ and $q = 2$. The constitutive dimerization constant, X^*, for the induced dimerization scheme is given by

$$X^* = C_{PS_2}/(C_M + C_{MS})^2 = Yk_{MS}^2C_S^2/(1 + k_{MS}C_S)^2 \tag{6.21}$$

whereas the corresponding expression for the system with preferential binding (eq. 6.18 with $n = 2$, $p = 1$, $q = 2$) is

$$X^* = X(1 + k_{PS}C_S)^2/(1 + k_{MS}C_S)^2 \tag{6.22}$$

Although eqs. 6.21 and 6.22 both predict systematic increases in the magnitude of X^* with increasing C_S, the forms of these dependences differ quantitatively (Nichol and Winzor, 1976). The expression for the ligand-induced associating system (eq. 6.21) may be rewritten as

$$(1/\sqrt{X^*}) = 1/(k_{MS}C_S\sqrt{Y}) + (1/\sqrt{Y}) \tag{6.23}$$

which signifies a linear dependence of $1/\sqrt{X}$ upon $1/C_S$. Similar treatment of eq. 6.22 leads to the expression

$$(1/\sqrt{X^*}) = (1 + k_{MS}C_S)/[(1 + k_{PS}C_S)\sqrt{X}] \tag{6.24}$$

which signifies a curvilinear dependence of $1/\sqrt{X^*}$ upon $1/C_S$ for a system with pre-existence of the dimerization.

In systems where sigmoidal effects are linked with changes in the self-association state of the acceptor, it is thus possible to distinguish between models based on ligand-induced and pre-existing self-association. However, the possibility should also be contemplated that sites within an individual state of a self-associating acceptor may exhibit cooperativity. Although expressions such as eqs. 6.13 and 6.18 may be extended to incorporate positive and negative cooperativity of sites (Saroff, 1991), the consequent increase in the number of parameters requiring evaluation usually precludes their use for characterization of the interplay of equilibria. That approach has been adopted in studies of the binding of competitive inhibitors to α-chymotrypsin (Nichol *et al.*, 1972a), because for that relatively simple system the only change required in the expression for X^* (eq. 6.22) is replacement of the $(1 + k_{PS}C_S)^2$ term by $[(1 + k_{PS}C_S(2 + k_{P'S}C_S)]$ to accommodate the possibility that ligand binding to a second site on the dimer may be governed by an altered binding constant ($k_{P'S}$). Consideration of the two sites on dimeric α-chymotrypsin to be equivalent and independent was thereby justified (Nichol *et al.*, 1972a).

Finally, it should be noted that the extension of the Monod model to include a pre-existing self-association of the acceptor can also be modified readily to include systems in which the pre-existing equilibrium involves dissimilar species ($A + B \rightleftarrows AB$). For example, B may be a cellular membrane to which A

binds (Masters *et al.*, 1969), or A and B may be enzymes catalyzing consecutive reactions of a pathway (Nichol *et al.*, 1974a). Preferential binding of ligand to either A or AB provides the necessary interplay of equilibria to bring about a sigmoidal response. Indeed, the only requirement for such a response to be obtained is an interplay of equilibria wherein one species coexists in two states. Inasmuch as the emphasis thus far has centered on systems in which that species is the acceptor, it is appropriate to conclude this chapter by considering situations in which the ligand is the species involved in the pre-existing association equilibrium.

6.6. EFFECTS OF LIGAND SELF-ASSOCIATION

In discussions of binding theory there is a tendency to regard the acceptor as the species undergoing self-association or isomerization and, therefore, to regard such self-interaction as the only potential source of allostery. However, there are also situations in which the reactant undergoing self-interaction is the ligand. The possibility of control effects arising from equilibrium coexistence of two ligand states assumes particular significance when viewed in the light that many ligands isomerize or self-associate. Examples include the isomerization of *N*-acetylglucosamine (Thomas, 1966) and the self-association of ATP (Ferguson *et al.*, 1974; Heyn and Bretz, 1965), organic dyes (Lamm and Neville, 1965), cholesterol (Parker and Bhaskar, 1968), purines (Van Holde and Rosetti, 1967), and pyrimidines (Ts'o *et al.*, 1963).

We consider, for illustrative purposes, a relatively simple situation in which each state of a self-associating ligand ($nL \leftrightarrows L'$) interacts competitively with the same p independent and equivalent sites on an acceptor A for which the respective intrinsic binding constants are k_{AL} and $k_{AL'}$. From experimental and biological viewpoints it is relevant to define the binding function as

$$r = (\overline{C}_S - C_L - nC_{L'})/\overline{C}_A \tag{6.25}$$

in which case the binding equation becomes (Nichol *et al.*, 1969)

$$r = p(k_{AL}C_L + nk_{AL'}C_{L'})/(1 + k_{AL}C_L + k_{AL'}C_{L'}) \tag{6.26}$$

A rectangular hyperbolic dependence of r upon $\overline{C}_L$ (*i.e.*, $C_L + nC_{L'}$), results from ligand isomerization ($n = 1$), whereas a sigmoidal response emanates from eq. 6.26 when $n > 1$ and $k_{AL} = 0$, the situation when the acceptor exhibits affinity solely for the polymeric form of the ligand. The curves assume a negatively cooperative form in the event that monomeric ligand is the exclusive form of ligand bound (Nichol *et al.*, 1969). The latter result has been obtained in studies of the binding of prothrombin fragment 1 to single-layer phospholipid vesicles (Dombrose *et al.*, 1979), a result in keeping with the hypothesis that interaction of the monomeric form of fragment 1 with vesicles is competitive with its dimerization.

An example of the situation in which the ligand polymer (L') binds preferentially (rather than exclusively) to the acceptor is provided by the interaction of chlorpromazine with tubulin (Hinman and Cann, 1976)—a system for which the Scatchard plot exhibits two critical points, a minimum and a maximum (Fig. 6.11a). Although the unusual form of this Scatchard plot was interpreted initially as signifying relatively strong interaction of chlorpromazine with a single tubulin site and weaker, positively cooperative binding of an additional eight or nine chlorpromazine molecules (Hinman and Cann, 1976), it has subsequently been noted that preferential binding of micellar chlorpromazine ($n = 9$–11) to a single tubulin site ($p = 1$, $k_{AL'} > k_{AL}$) is also a plausible explanation of the results (Sculley *et al.*, 1981). The latter interpretation has been supported by the frontal gel chromatographic demonstration (Fig. 6.11b) that chlorpromazine undergoes micellization in the appropriate concentration range for this interpretation to be applicable (Cann *et al.*, 1981).

In binding studies with relatively hydrophobic ligands the self-association status of the ligand assumes particular importance because of the tendency of amphipathic ligands to form micelles. The above frontal gel chromatographic investigation of chlorpromazine (Fig. 6.11b) provides one example of the determination of the self-association characteristics of a ligand, the same ligand having also been subjected to sedimentation equilibrium in order to characterize chlorpromazine self-association under a different set of conditions (Nichol *et al.*, 1982). In a recent study exploiting the micellar state of the ligand, advan-

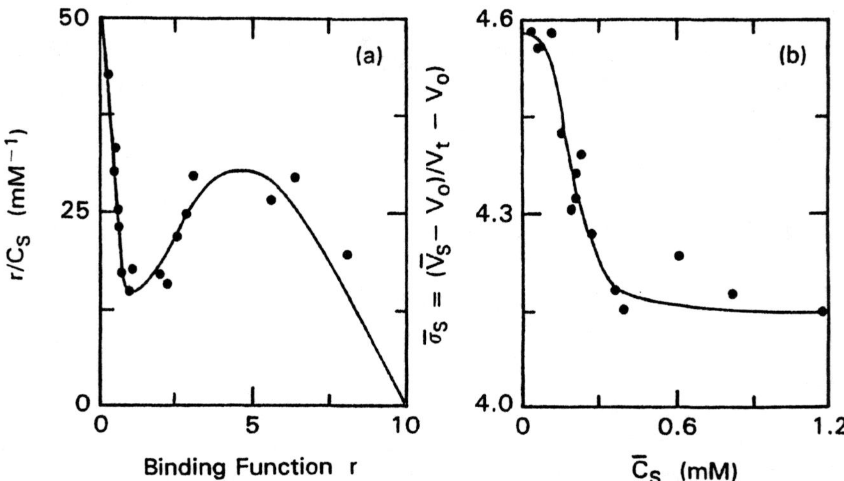

Fig. 6.11. Preferential binding of the micellar state of chlorpromazine to brain tubulin. **(a)** Scatchard plot of the binding data. [Adapted from Hinman and Cann (1976).] **(b)** Characterization of chlorpromazine micellization by frontal gel chromatography on Sephadex G-25: $\bar{\sigma}_S$ is the partition coefficient corresponding to elution volume $\bar{V}_S$. [Adapted from Cann *et al.* (1981).]

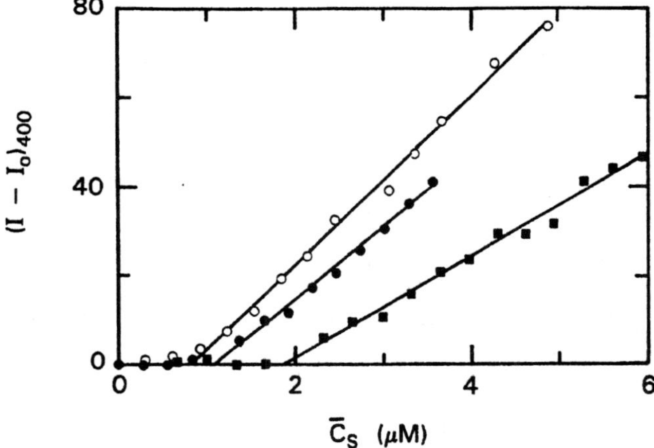

Fig. 6.12. The effect of lipoprotein lipase (*A*) on the apparent critical micelle concentration of oleic acid (*S*), obtained as the break point in the dependence of light scattering at 400 nm upon concentration of fatty acid in experiments with zero (○), 0.08 μM (●), and 0.30 μM (■) lipoprotein lipase. [Adapted from Edwards *et al.* (1994).]

tage has been taken of the increase in apparent critical micelle concentration of fatty acids (Fig. 6.12) to characterize the exclusive interaction of monomeric fatty acids with lipoprotein lipase (Edwards *et al.*, 1994).

Situations may also be encountered in which the ligand interaction competitive with binding to the acceptor is not confined to self-association. For example, Denburg and De Luca (1968) have shown that firefly luciferase exhibits Michaelis-Menten kinetic behavior when magnesium pyrophosphate is used as substrate. However, a decreased level of Mg^{2+} results in sigmoidal kinetic behavior, a result commensurate with the observation (Nichol *et al.*, 1969) that a sigmoidal response arises in situations where the complexed form of ligand is binding preferentially to the acceptor.

In summary, this chapter has served to emphasize the diversity of interactions that can give rise to sigmoidal binding responses. Such responses may frequently be considered to reflect an interplay of equilibria involving competition between two equilibrium reactions involving the acceptor: the first being the interaction of the acceptor with the ligand and the second being the interaction of the acceptor with a different macromolecular state of itself (conformational isomer or oligomer)—Figures 6.5, 6.6, and 6.8. In other systems, however, a sigmoidal binding curve may reflect the equilibrium coexistence of ligand in two states, the complexed form of which interacts preferentially with acceptor (Fig. 6.11). However, despite this diversity of reaction schemes that can give rise to a sigmoidal binding response, this chapter also offers cause for confidence that the performance of a carefully constructed set of experiments can at least identify the particular interplay of equilibria responsible for the sigmoidal effect and, under favorable circumstances, lead to a quantitative thermodynamic characterization of the interactions.

7

COMPLICATING FACTORS IN THE ANALYSIS OF BINDING CURVES

In the previous chapter we have discussed ways in which the deviation of a binding response from rectangular hyperbolic dependence upon C_S may be used in conjunction with other measurements to identify and in many cases to characterize the interplay of equilibria responsible for the form of the binding curve. However, there are often additional factors that need to be taken into account before such analysis is attempted. The effect of ligand multivalence on the form of the binding curve clearly needs to be addressed in view of the fact that many ligands of interest are proteins possessing more than one site for attachment to their specific acceptor (receptor), and the effect of ligand partitioning into membranes is obviously pertinent to studies of interactions between hydrophobic ligands and membrane receptors. The aim of this chapter is to address the consequences of these phenomena as well as those of thermodynamic nonideality on interactions in a crowded molecular environment.

7.1. ALLOWANCE FOR LIGAND MULTIVALENCE

All binding equations presented thus far have been derived on the basis that the ligand is univalent and therefore should not be used to interpret the binding behavior of systems in which the ligand is multivalent. This limitation of the binding expressions has seldom been appreciated, because there are countless examples of binding studies in which the Scatchard plot has been used to analyze, for example, the interactions of antibodies with their specific surface antigens. The problem with that course of action is that ligand multivalence (bivalence in the case of an immunoglobulin G) introduces curvilinearity into the Scatchard plot of the same form as that for binding of a univalent ligand to nonequivalent or negatively cooperative acceptor sites (Calvert *et al.*, 1979). Allowance for ligand multivalence in the analysis of binding data was first made in the context of quantitative affinity chromatography (Nichol *et al.*, 1981b; Winzor *et al.*, 1982), and, indeed, a general counterpart of the Scatchard plot (Hogg and Winzor, 1985) has evolved from the linear transformation applied to the quantitative affinity chromatography expressions (Hogg and Winzor, 1984).

7.1.1. The Multivalent Scatchard Formulation

Consider a situation in which a ligand, S, with f sites binds to an acceptor, A, with p sites for ligand. Because of the ligand multivalence, the binding function needs to be calculated as

$$r_f = (\overline{C}_S^{1/f} - C_S^{1/f})/\overline{C}_A \tag{7.1}$$

where C_S denotes the molar concentration of free ligand in a mixture with $\overline{C}_S$ and $\overline{C}_A$ as the respective total molar concentrations of ligand and acceptor (Hogg and Winzor, 1985). As required, this expression becomes eq. 1.3 (the traditional binding function, r) if f is set equal to unity. In terms of this binding function, the multivalent counterpart of the Scatchard analysis for equivalent and independent binding governed by intrinsic association constant, k_{AS}, is

$$r_f/C_S^{1/f} = pk_{AS} - fk_{AS}r_f\overline{C}_S^{(f-1)/f} \tag{7.2}$$

A linear plot of $r_f/C_S^{1/f}$ *versus* $r_f\overline{C}_S^{(f-1)/f}$ is thus the requirement for equivalence and independence of sites for a ligand that is multivalent. Those familiar with traditional Scatchard analysis of binding data may at first query the presence of a term in total ligand concentration within the abscissa parameter. However, on setting $f = 1$ in eq. 7.2, the term becomes unity by virtue of the power (zero) to which $\overline{C}_S$ is raised when the ligand is univalent.

An obvious prerequisite for the application of eqs. 7.1 and 7.2 to binding data is the assignment of a magnitude to the ligand valence, f. Although a degree of reluctance about assigning this value is certainly understandable, it must be clearly understood that any attempt to escape the issue by resorting to traditional Scatchard analysis merely means that a value of unity has unwittingly been chosen as the most appropriate valence of the ligand. To emphasize this point we consider the results from a partition equilibrium study of the binding of aldolase to muscle myofibrils (Kuter *et al.*, 1983).

Figure 7.1a presents a plot of those results in accordance with conventional Scatchard analysis, a course of action that has, as stressed above, assigned a value of unity to f, the valence of aldolase. Any attempted quantitative interpretation of the curvilinearity would presumably be in terms of myofibrillar sites that exhibit heterogeneity or negative cooperativity in their interactions

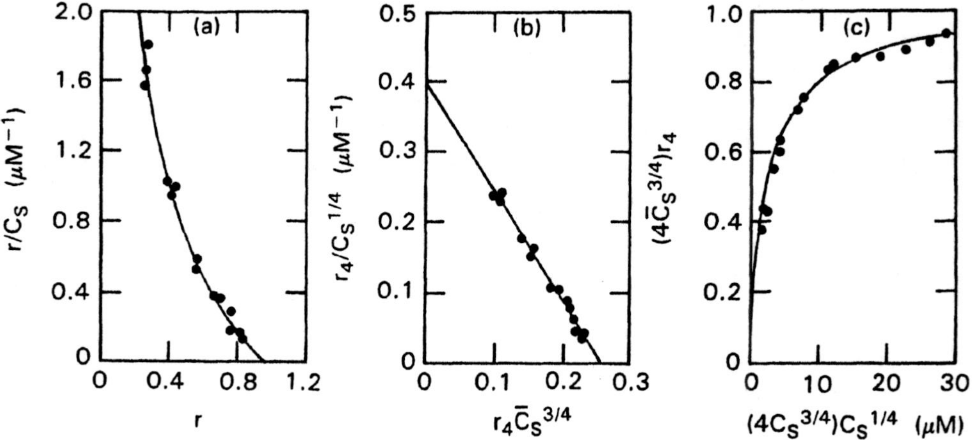

Fig. 7.1. Analysis of binding data for the interaction of aldolase with rabbit muscle myofibrils, the results being taken from Table 1 of Kuter *et al.* (1983). (**a**) Conventional Scatchard plot. (**b**) Multivalent Scatchard plot according to eq. 7.2 with $p = 4$. (**c**) Rectangular hyperbolic plot in accordance with eq. 7.3 for a tetravalent ligand. [Adapted from Harris *et al.* (1995a).]

with the enzyme. However, aldolase is known to be a tetrameric enzyme with four equivalent and independent active sites (Lai and Horecker, 1972); and the involvement of the active-site region in the adsorption of enzyme is implicated by the demonstration that the interaction of aldolase with the myofibrillar matrix is competitive with enzyme catalysis (Harris and Winzor, 1987). Reanalysis of the results in terms of the generalized Scatchard formulation (eq. 7.2 with $f = 4$) is presented in Figure 7.1b. The linearity of this plot and its conformity with the mandatory abscissa intercept of 0.25 ($1/f$) establishes the adequacy of a single intrinsic association constant to describe the binding of this tetravalent ligand to the myofibrillar matrix. Linear regression analysis of the experimental results leads to the conclusion that $p = 1.02$ (± 0.02) and $k_{AS} = 3.8$ (± 0.3) $\times 10^5$ M^{-1}. In these experiments the total concentration of acceptor, $\bar{C}_A$, used to calculate r_f was taken as the myofibrillar capacity for aldolase, which corresponds to the concentration of binding sites for aldolase. The value for p that emanated from the analysis is thereby explained.

The above graphical approach to the evaluation of the binding constant for a multivalent ligand is open to criticism because of its reliance upon transformed experimental parameters and the consequent statistically distorted data distribution that results from such transformation (Thompson and Klotz, 1971; Klotz, 1983). Such action can now be avoided by rewriting eqs. 7.2 in the form (Harris *et al.*, 1995a)

$$fr_f\bar{C}_S^{(f-1)/f} = pf\bar{C}_S^{(f-1)/f}k_{AS}C_S^{1/f}/(1 + f\bar{C}_S^{(f-1)/f}k_{AS}C_S^{1/f}) \qquad (7.3)$$

Because the dependence of $fr_f\overline{C}_S^{(f-1)/f}$ upon $f\overline{C}_S^{(f-1)/f}C_S^{1/f}$ is rectangular hyperbolic, p and k_{AS} may be obtained by substituting $fr_f\overline{C}_S^{(f-1)}$ for r and $f\overline{C}_S^{(f-1)/f}C_S^{1/f}$ for C_S in standard computer programs for the evaluation of binding (or kinetic) parameters. The rectangular hyperbolic form of the suggested plot for the aldolase myofibrils is confirmed in Figure 7.1c, which yields estimates of p and k_{AS} that are indistinguishable from those obtained by analysis of Figure 7.1b.

If the total concentration of acceptor, $\overline{C}_A$, is of unknown but constant magnitude in acceptor–ligand mixtures, eq. 7.3 may be multiplied throughout by $\overline{C}_A$ to allow advantage to be taken of the fact that $r_f\overline{C}_A = (\overline{C}_S^{1/f} - C_S^{1/f})$. Nonlinear regression analysis on the basis that the dependence of $f\overline{C}_S^{(f-1)/f}(\overline{C}_S^{1/f} - C_S^{1/f})$ upon $f\overline{C}_S^{(f-1)/f}C_S^{1/f}$ is rectangular hyperbolic then yields k_{AS} and $p\overline{C}_A$, the total concentration of acceptor sites (Harris *et al.*, 1995a).

The availability of the rectangular hyperbolic binding equation for multivalent ligands (as well as its linear Scatchard transform) now provides an opportunity for critical appraisal of the interpretations placed on results for the interactions of many macromolecular ligands with cellular receptors—a field where univalence of ligands such as antibodies and proteins with quaternary structure has been a prevalent, unwitting, and dubious assumption.

7.1.2 Consequences of Cooperativity

In Chapter 6 guidelines were presented for identifying heterogeneity and cooperativity in the binding of univalent ligands to acceptors. Multivalence of the ligand invalidates those criteria for heterogeneity and cooperativity of binding sites, because the simple rectangular hyperbolic relationship ceases to describe the binding curve for acceptor–ligand interactions governed by a single intrinsic binding constant. Section 7.1.1 established the criteria for identifying the adequacy of a single intrinsic association constant to describe the binding of a multivalent ligand; and hence it is now appropriate to consider the form that the binding curve assumes under conditions deviating from that simplest model of binding behavior—a problem that has not been addressed until recently (Harris *et al.*, 1995b).

The major outcome of those deliberations is that multivalence of the ligand necessitates the consideration of an additional source of cooperativity inasmuch as any deviation from rectangular hyperbolic form of the binding curve, $fr_f\overline{C}_S^{(f-1)/f}$ *versus* $f\overline{C}_S^{(f-1)/f}C_S^{1/f}$, may reflect cooperativity (or heterogeneity) of ligand sites rather than of acceptor sites. For systems in which the cooperativity resides in the acceptor sites, the criteria used for the interpretation of binding curves for univalent ligands also apply to data for multivalent ligands provided that the results are analyzed in terms of the rectangular hyperbolic relationship for an f-valent ligand (eq. 7.3). Consequently, if acceptor sites are the source of cooperative binding, multivalent Scatchard plots of the forms shown in Figure 7.2a,b signify the existence of positive and negative cooperativity, respectively. The latter plot (Fig. 7.2b) could, of course, indicate heterogeneity rather than negative cooperativity, as in the univalent case. On the other hand, if it is assumed that cooperativity exists between the sites of the multivalent ligand, a different set of criteria must be used to identify the positive and negative forms of that cooperativity. Thus, a multivalent Scatchard plot of the form shown in Figure 7.2a becomes symptomatic of negatively cooperative ligand sites, the alternative form of curvilinearity (Fig. 7.2b) being the symptom of ligand sites that are positively cooperative. In an experimental context a decision therefore needs to be made whether a binding curve of the form shown in Figure 7.2b should be interpreted in terms of negatively cooperative acceptor sites or of positively cooperative sites on the multivalent ligand. Clearly, additional information about the acceptor and the ligand is required to guide that decision. In that regard, results obtained for the binding of pyruvate kinase to muscle myofibrils have been used to illustrate the type of approach that may be used to sort out this dilemma (Harris *et al.*, 1995b).

In summary, the multivalent counterpart of the Scatchard plot assumes the role previously assigned to the original Scatchard analysis for the identification of cooperative binding. However, because conflicting criteria apply to cooperativity emanating from acceptor and ligand sites, a decision needs to be made about the likely source of the effect before the form of the curvilinear plot can be interpreted. At this stage no general guidelines can be given on ways to facilitate that decision. Too little time has elapsed since realization of the existence of the dilemma.

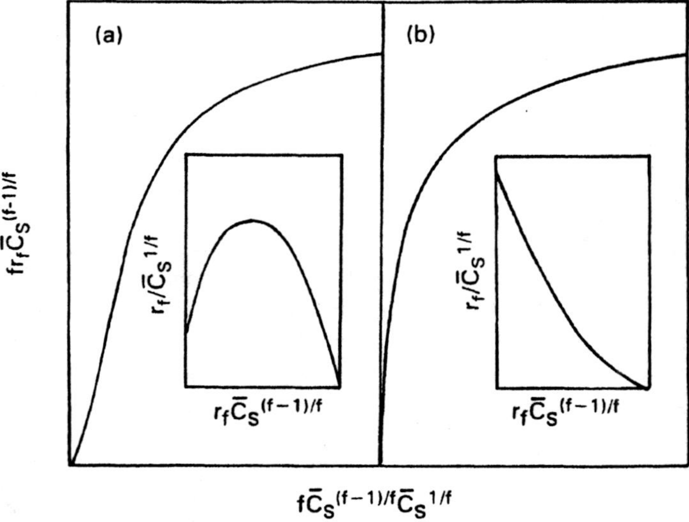

Fig. 7.2. Conflicting criteria for cooperative binding of multivalent ligands. (**a**) Schematic multivalent plots signifying either positive cooperativity of acceptor sites or negative cooperativity of ligand sites. (**b**) Corresponding plots signifying either negative cooperativity of acceptor sites or positive cooperativity of ligand sites.

7.1.3. Precipitin Effects

In situations where the acceptor and ligand are both multivalent, there is clearly the possibility of forming an array of complexes comprising three-dimensional networks of alternating A and S units. Indeed, such formation of vast cross-linked networks has long provided the conceptual basis for interpreting precipitin effects with multivalent antigens and bivalent antibodies (Heidelberger and Kendall, 1935; Goldberg, 1952, 1953). Theoretical expressions have been developed for the situation in which the bivalent antibody, A, and the f-valent antigen (S) bear functional groups of equal reactivity, whereupon a single intrinsic binding constant, k_{AS}, governs all equilibria. A striking feature of the binding behavior of such systems is the occurrence of a maximal extent of reaction at $(C_A)_m = 1/2k_{AS}$, this being a concentration at which maximal precipitation is likely to be observed (Goldberg, 1952, 1953; Singer, 1965; Calvert *et al.*, 1979). Thus, determination of the free concentration of antibody associated with maximal precipitation affords, in principle, a simple method of evaluating the intrinsic binding constant.

Because $(\overline{C}_A)_m$, the total antibody concentration associated with maximal precipitation, is a more readily available experimental parameter, an alternative expression (Nichol *et al.*, 1982),

$$(1/k_{AS}) = 2(\overline{C}_A)_m - f\overline{C}_S \tag{7.4}$$

offers a better prospect for determining k_{AS}. The use of eq. 7.4 for evaluating k_{AS} is, of course, subject to the availability of a value for f, the number of sites for antibody on the antigen. Theoretically, this value is obtained as the stoichiometry of the antigen–antibody complex in the limit of infinite antibody concentration (Goldberg, 1952, 1953; Nichol *et al.*, 1982). Experimentally, f has been taken as the molar antibody/antigen ratio obtained by extrapolating a Heidelberger (1939) plot (the molar antibody/antigen ratio of the precipitate as a function of antigen concentration in the precipitate) to the ordinate intercept on the grounds that such a value describes the limiting composition of the antigen–antibody precipitate in the region of antibody excess.

The use of eq. 7.4 to determine k_{AS} is illustrated in Figure 7.3, which summarizes results for the reactions of the α_m and α_c forms of α-crystallin with an elicited monoclonal antibody (Winzor *et al.*, 1989). The immunoprecipitation behavior is described reasonably by a common asymmetric curve (Fig. 7.3a) with maximal precipitation at 2.1–2.3 mg/ml antibody: $(\overline{C}_A)_m$ is thus 13–14 μM for this immunoglobulin G. The corresponding Heidelberger plots of the precipitate composition are shown in Figure 7.3b, which presents results from the region of antibody excess ($\overline{C}_A > \overline{C}_S$). Extrapolation of these results to zero precipitation yields estimates of 8 and 16 for the respective antibody/antigen ratios in the limiting precipitates obtained with α_m- and α_c-crystallins, respectively. Substitution of these valences (f), which correspond to one antibody molecule per dimer of α-crystallin subunits, into eq. 7.4 yields an

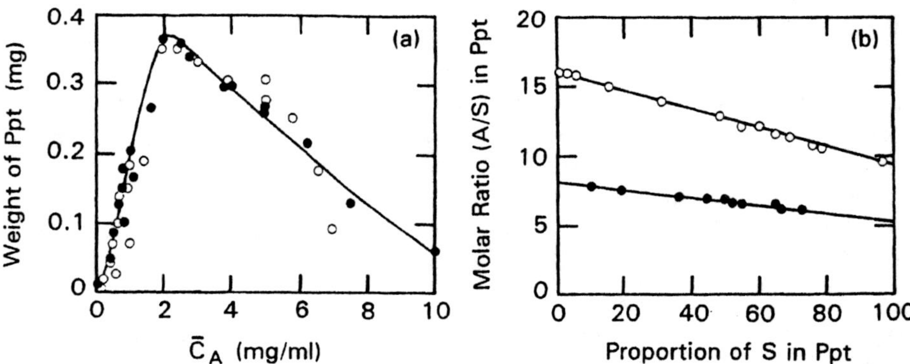

Fig. 7.3. Use of a precipitin curve to evaluate the binding constant for the interactions of α_m (●) and α_c (○) crystallins with an elicited monoclonal antibody. (**a**) Determination of $(\overline{C}_A)_m$, the total antibody concentration for maximal precipitation in mixtures containing 0.63–0.64 mg/ml antigen. (**b**) Heidelberger plot of the precipitate composition to determine the antigen valence (f) required to evaluate k_{AS} via eq. 7.4.

intrinsic binding constant of 9 ($\pm$2) $\times$ 10^4 M^{-1} for the interaction of either form of α-crystallin with the monoclonal antibody.

The value of k_{AS} so obtained needs to be used to check on the extent of antigen saturation prevailing in the range of antibody concentrations relevant to extrapolation of the Heidelberger plot. For that purpose advantage is taken of the expression (Calvert *et al.*, 1979)

$$P_S = 1 + [(1 + 2k_{AS}\overline{C}_A - fk_{AS}\overline{C}_S - \sqrt{\Delta})/(2fk_{AS}\overline{C}_S)] \qquad (7.5a)$$

$$\Delta = (1 + 2k_{AS}\overline{C}_A - f\overline{C}_S)^2 + 4fk_{AS}\overline{C}_S \qquad (7.5b)$$

where P_S denotes the probability that an antigen site has reacted with antibody. For the above system P_S was in the vicinity of 0.92–0.96 (Winzor *et al.*, 1989) and thus approached its limiting value of unity for much of the antibody concentration range covered by the Heidelberger plot. Justification is thereby provided for the assumption that extrapolation of the antigen/antibody ratio for the precipitate to zero precipitation yields the stoichiometry of the complex formed in the limit of infinite antibody concentration (A_fS).

The availability of a procedure to determine the intrinsic binding constant from the precipitin curve for an antibody–antigen interaction is an important advance in the quantitative characterization of immunochemical responses involving multivalent antigens. These reactions cannot be studied by many other methods because of the cross-linking and precipitation that occurs in such systems. That problem feature is used to advantage in the above procedure, which illustrates an unrealized potential of this long-established immunoprecipitation technique.

7.1.4. Quantitative Affinity Chromatography

By establishing the basic procedure for accommodating multivalence in binding equations (eqs. 7.2 and 7.3), we are now in a position to make similar amendments to the quantitative expressions used for the characterization of ligand interactions by competitive binding methods. In quantitative affinity chromatography the partition of a multivalent solute is described by the expression (*cf.* eq. 5.2)

$$(\bar{\bar{C}}_A^{1/p} - \bar{C}_A^{1/p})/\bar{C}_A^{1/p} = k_{AX}\bar{\bar{C}}_X - pk_{AX}\bar{C}_A^{(p-1)/p}(\bar{\bar{C}}_A^{1/p} - \bar{C}_A^{1/p}) \qquad (7.6)$$

where, as in Chapter 5, $\bar{C}_A$ is the measured concentration of partitioning solute in the liquid phase of a system with total solute concentration $\bar{\bar{C}}_A$; and where k_{AX}, the intrinsic binding constant for the solute–matrix interaction, is replaced by its constitutive counterpart, $\bar{k}_{AX} = k_{AX}/(1 + k_{AI}C_I)$, in experiments where the liquid phase is supplemented with a free concentration, C_I, of ligand that competes with matrix sites for the partitioning solute. It is important to recognize that the use of eq. 5.3 to inter-relate k_{AX} and $\bar{k}_{AX}$ implies univalence of the competing ligand. This is also a necessary experimental restriction in experiments with a multivalent partitioning solute to prevent removal of *A* from the liquid phase through the formation of cross-linked networks on the matrix (see Section 7.1.3): eq. 7.6 only takes into account the partitioning of *A* by direct interaction with matrix sites. On the grounds that either *A* or *I* must therefore be univalent, multivalence of the competing ligand can only be accommodated in quantitative affinity chromatography (and other competitive binding methods) if the partitioning solute is univalent. In those circumstances C_I in eq. 5.3 is replaced by fC_I, the concentration of competing ligand sites (Winzor and Jackson, 1993; Ward *et al.*, 1995).

In column chromatography experiments with a multivalent partitioning solute the counterpart of eq. 7.6 (and of eq. 5.6 for a univalent solute) is

$$[(\bar{V}_A/V_A^*)^{1/p} - 1] = k_{AX}\bar{\bar{C}}_X - p(\bar{V}_A/V_A^*)^{(p-1)/p}\bar{C}_A[(\bar{V}_A/V_A^*)^{1/p} - 1] \qquad (7.7)$$

in which k_{AX} is again replaced by $\bar{k}_{AX}$ to take into account the presence of a univalent competing ligand, *I*.

The application of these expressions is illustrated in Figure 7.4, where eq. 7.6 is used to analyze partition equilibrium data (Kuter *et al.*, 1983) for the effect of phosphate on the interaction of aldolase with muscle myofibrils (Fig. 7.4a), and where eq. 7.7 is the expression for characterizing the interaction of NADH with lactate dehydrogenase on the basis of its effect on the elution of the enzyme from a column of trinitrophenyl-Sepharose (Fig. 7.4b). The partitioning enzyme is tetravalent ($p = 4$) in both systems. From the effects of the concentration of competing ligand on $\bar{k}_{AX}$ (insets in Fig. 7.4), a value of

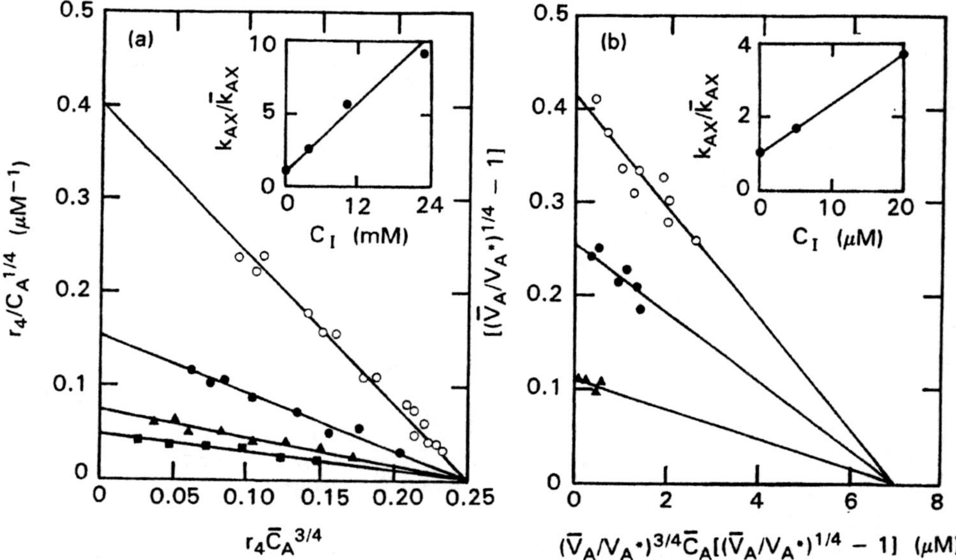

Fig. 7.4. Evaluation of binding constants for multivalent partitioning solutes by quantitative affinity chromatography. **(a)** Results, plotted in accordance with eq. 7.5, of partition equilibrium studies (Kuter *et al.*, 1983) of the interaction of aldolase with myofibrils in the presence of zero (○), 4 mM (●), 10 mM (▲), and 23 mM (■) phosphate. **(b)** Results, plotted in accordance with eq. 7.6, of a frontal chromatographic study of lactate dehydrogenase on a column of trinitrophenyl–Sepharose equilibrated with zero (○), 5 μM (●), and 20 μM (▲) NADH. [Adapted from Bergman and Winzor (1986).] Insets represent the plots of the resultant $\bar{k}_{AX}$values to obtain k_{AI} via eq. 5.3a.

400 M^{-1} is obtained for the binding of phosphate to aldolase (Kuter *et al.*, 1983), whereas k_{AI} for the interaction of NADH with rabbit muscle lactate dehydrogenase is calculated to be 1.3×10^5 M^{-1} (Bergman and Winzor, 1986).

7.1.5. Competitive Radioimmunoassays

Allowance for multivalence of an antigen is also readily incorporated into the basic expression for its reaction with antibody in radioimmunoassays. In situations where the same binding constant (k_{AS}) applies to the interactions of antigen (S) and radiolabeled antigen (S^*), the general analog of eq. 5.18 for the distribution of labeled antigen (total concentration $\bar{C}_{S^*}$) in the presence of cold antigen is (Winzor *et al.*, 1991a,b)

$$(1 - \alpha^{1/f})/\alpha^{1/f} - (1 - \alpha_s^{1/f})/\alpha_s^{1/f}$$
$$= fk_{AS}[(1 - \alpha_s^{1/f})(\bar{C}_S + \bar{C}_{S^*}) - (1 - \alpha^{1/f})\bar{C}_{S^*}] \qquad (7.8)$$

where α denotes the free fraction of f-valent labeled antigen in the absence of cold antigen, and α_s denotes the corresponding parameter in the presence of a concentration $\overline{C}_S$ of unlabeled antigen. Because k_{AS} is the only parameter of unknown magnitude in eq. 7.7, the intrinsic binding constant for the antibody–antigen interaction may be estimated by regression analysis of the results in terms of this expression for any assigned value of f.

Equation 7.8 draws attention to the assumption inherent in the Müller (1980, 1983) method that the antigen is univalent, because the selection of $\overline{C}_{S^*}$ and $\overline{C}_S$ as the concentrations for 50% and 25% bound radiolabel, respectively, is now seen to entail the supposition that $(1 - \alpha^{1/f}) = 0.5$ and $(1 - \alpha_s^{1/f}) = 0.25$. If eqs. 5.19 and 5.20 are to be applied to results for an f-valent antigen and its analogs, the free fractions required to retain their validity are $\alpha^{1/f} = 0.5$ and $\alpha_s^{1/f} = 0.75$. The determination of k_{AS} (or k_{AI}) for a bivalent antigen would require both the selection of $\overline{C}_{S^*}$ such that 75% of the radiolabeled antigen was bound in the absence of cold antigen, and the determination of the unlabeled antigen concentration ($\overline{C}_S$) that decreases this proportion to 44%.

In the presence of a total concentration $\overline{C}_I$ of an f-valent competitive inhibitor of radiolabeled antigen binding, the distribution of labeled ligand is related to the inhibitor constant, k_{AI}, by the multivalent analogs of eqs. 5.13 and 5.15, namely,

$$k_{AI} = (1 - \beta)f\overline{C}_I k_{AS}\alpha_i^{1/f}[(1 - \alpha_i^{1/f})\beta f\overline{C}_I] \tag{7.9}$$

$$(1 - \beta)f\overline{C}_I = f\overline{C}_S(\alpha_i^{1/f} - \alpha^{1/f}) + (1/k_{AS})[(1 - \alpha^{1/f})/\alpha^{1/f} - (1 - \alpha_i^{1/f})/\alpha_i^{1/f}] \tag{7.10}$$

which may be used to evaluate k_{AI} as the only parameter of unknown magnitude.

Figure 7.5 summarizes the analysis of radioimmunoassay data for the binding of [^{125}I]fibrinogen (S^*) and some fibrinopeptides (I) to an anti-fibrinogen

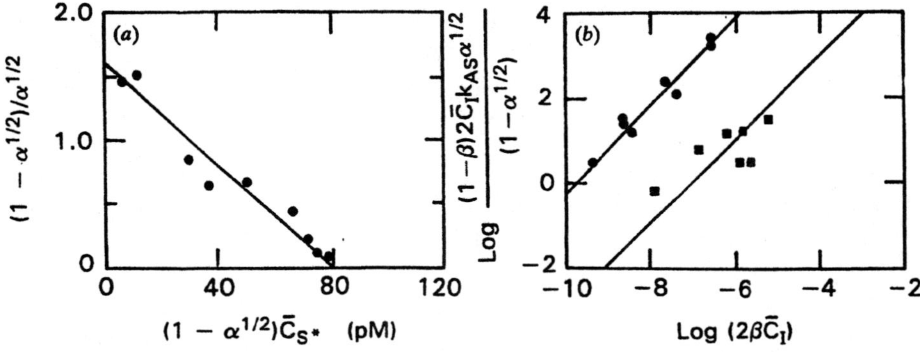

Fig. 7.5. Allowance for multivalence of the antigen in the characterization of interactions by radioimmunoassay. (a) Multivalent Scatchard plot to evaluate k_{AS} for the interaction of [^{125}I] fibrinogen, a bivalent antigen, with an elicited antibody preparation. (b) Application of eqs. 7.8 and 7.9 to obtain binding constants (k_{AI}) for the competitive inhibition of [^{125}I] fibrinogen binding by native fibrinogen (●) and the disulfide knot fragment derived therefrom (■). [Adapted from Winzor *et al.* (1991b).]

antibody preparation that had been purified by affinity chromatography on a column with a peptide comprising residues 16–25 of the α-chain of fibrinogen as immobilized ligand (Winzor *et al.*, 1991b). The evaluation of k_{AS} by multivalent Scatchard analysis (eq. 7.2) of results for mixtures of antibody (A) and radiolabeled fibrinogen (S^*) is presented in Figure 7.5a, whereas the evaluation of k_{AI} by the application of eqs. 7.8 and 7.9 to mixtures supplemented with I is summarized in Figure 7.5b. Of interest in that regard is the finding that consideration of native fibrinogen to be a competitive inhibitor of [^{125}I]fibrinogen binding leads to a value of k_{AI}, 8.1 ($\pm$3.0) $\times$ 10^9 M^{-1}, that essentially duplicates the k_{AS} of 1.0 ($\pm$0.1) $\times$ 10^{10} M^{-1} deduced from Figure 7.5a. Understandably, the interaction of the disulfide knot fragment of fibrinogen (also bivalent) exhibits a weaker interaction (k_{AI} = 2.1 $\times$ 10^7 M^{-1}) with the antibody, which was an elicited response to the whole fibrinogen molecule.

From the present viewpoint, the most important feature of Figure 7.5 is the finding that the antibody preparation recognizes the antigen (S) and its radiolabeled derivative (S^*) equally well. Such demonstration of agreement between k_{AS} and k_{AI} would clearly have validated the inherent assumption in the analysis of Figure 7.5a if S^* had been regarded as a tracer for the distribution of cold S. However, it also emphasizes that this questionable assumption for some systems (De Leo and Helmerhorst, 1992) can be avoided altogether by regarding the native antigen as a competitive inhibitor of S^* (see Section 5.3.2).

7.2 UNJUSTIFIED USE OF TOTAL FOR FREE LIGAND CONCENTRATION

Possibly as the result of familiarity with enzyme kinetic practice, there is a persistent tendency to employ $\overline{C}_S$ rather than C_S in the abscissa parameter of binding curves. Because the binding equation is based on the law of mass action, the required ligand concentration is C_S. In enzyme kinetics too the abscissa of the rectangular hyperbolic dependence is C_S, but the enzyme concentration is sufficiently small in relation to substrate (ligand) concentration ($\overline{C}_S >> p\overline{C}_A$) to permit the approximation that $C_S \approx \overline{C}_S$. This is sometimes a reasonable approximation in binding studies, particularly in those where the concentration of complexed ligand is being monitored; but in others it is not. Clearly, the approximation is totally unjustified in studies where binding is being monitored on the basis of the difference between total and free ligand concentrations.

In situations where $\overline{C}_S$ has been substituted unjustifiably for C_S, the Scatchard plot for a univalent ligand assumes the form symptomatic of positive cooperativity (see Fig. 3.5) for the simple reason that the binding function (or a parameter directly related thereto) is being divided by $\overline{C}_S$ instead of C_S to obtain the ordinate parameter. Because $C_S < \overline{C}_S$, the use of the total ligand concentration as an inappropriate estimate of C_S leads to underestimation of r/C_S, particularly in the range of low ligand concentration (small r). A curvilinear

Scatchard plot of positively cooperative form is the consequence. A similar situation also pertains to the analysis of data for multivalent ligands in terms of the general counterpart of the Scatchard plot.

Before assigning significance to deviations from rectangular hyperbolic behavior to positive cooperativity of acceptor sites (or possibly to negative cooperativity of multivalent ligand sites), it is obviously necessary to examine critically the validity of substituting $\overline{C}_S$ for C_S, which is only allowed if $\overline{C}_S >> p\overline{C}_A$. In that regard the above warning is not raised with the intention of trying to eliminate the use of the approximation. There are many circumstances under which it is totally justified. The important point is to realize that the assumption $C_S \approx \overline{C}_S$ has been made and that interpretation of the form of the binding curve (or a linear transform thereof) should be delayed until adequate consideration has been given to the acceptability of the approximation.

7.3. LIGAND PARTITIONING IN MEMBRANE-RECEPTOR SYSTEMS

When an acceptor protein is located in a particulate phase such as a lipid bilayer, and the ligand is effectively soluble only in the aqueous phase, classical analysis of the binding data still applies. Indeed, the binding study is simplified, since the membranes can usually be separated by centrifugation, thereby allowing the concentration of free ligand to be determined by analysis of the supernatant (Section 2.3.2). However, additional considerations apply when an amphipathic or hydrophobic ligand partitions into the membrane phase because of its compatibility with a nonaqueous environment. This section describes the characteristics of binding and partition processes, criteria that may be used for their distinction, and analytical procedures for the characterization of ligand binding when the two processes occur simultaneously.

7.3.1. Distinction Between Binding and Partition Processes

Binding processes are characterized by a rectangular hyperbolic dependence upon free ligand concentration if sites are equivalent and independent, and therefore have a binding capacity that is limited by the concentration of available acceptor sites. On the other hand, partition processes are nonsaturable, being limited only by the solubility of the ligand in each phase. They are described by a distribution or partition coefficient (Section 2.3.1) given by the ratio of the concentrations of ligand in the aqueous (α) and lipid (β) phases ($K_p = C_S^\beta/C_S^\alpha$). If the uptake of ligand by the membrane phase is regarded in binding terms, a plot of the binding function (r) *versus* the concentration of ligand in the aqueous phase (C_S^α) is linear, with a slope representing the partition coefficient. The corresponding Scatchard plot is horizontal. These features of ligand partitioning are evident in Figure 7.6, which refers to the uptake of metoprolol by human liver microsomes (Bogoyevitch *et al.*, 1987). An exper-

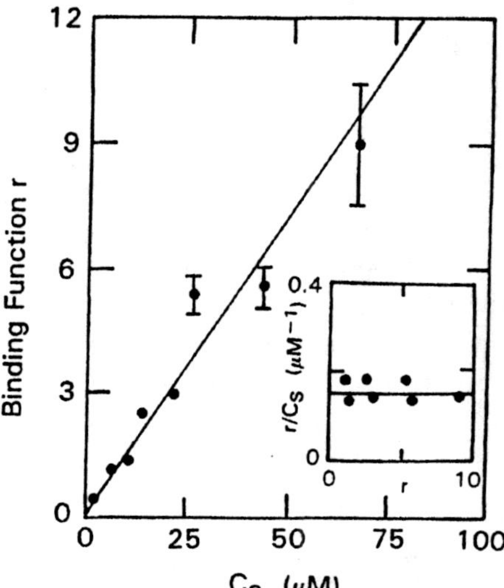

Fig. 7.6. Dependence of the binding function upon free ligand concentration for the uptake of metoprolol by human liver microsomes. [Adapted from Bogoyevitch *et al.* (1987).] Inset: Scatchard plot of the same dependence.

imenter should recognize the intrinsic characteristics of binding and partition processes in order to avoid interpreting one for the other. For example, limiting a binding study to values of C_S^α well below the dissociation constant ($1/k_{AS}$), where the binding curve is effectively linear, could lead to the incorrect conclusion that ligand uptake reflected partition rather than binding. In practice the identification of a partition process can often be difficult because the solubility limit of the ligand in either phase may not appear as a well-defined breakpoint in the plot of r *versus* C_S^α.

7.3.2. Simultaneous Binding and Partitioning in Membrane Systems

Two situations arise when the ligand binds to a membrane receptor and partitions into the lipid phase simultaneously but independently (Fig. 7.7). The binding site on the acceptor may be oriented toward the aqueous phase so that the relevant concentration of the free ligand is that in the aqueous phase (C_S^α). Under those conditions the expression for the binding function is simply

$$r = K_p C_S^\alpha + pk_{AS}C_S^\alpha/(1 + k_{AS}C_S^\alpha) \tag{7.11}$$

Alternatively, the binding site may be oriented toward the lipid phase, in which case the bound ligand is in direct equilibrium with the free ligand dissolved in the lipid phase (C_S^β), whereupon the description of the binding function in terms of the free ligand concentration in the aqueous phase becomes

$$r = K_p C_S^\alpha + pk_{AS}K_p C_S^\alpha/(1 + k_{AS}K_p C_S^\alpha) \tag{7.12}$$

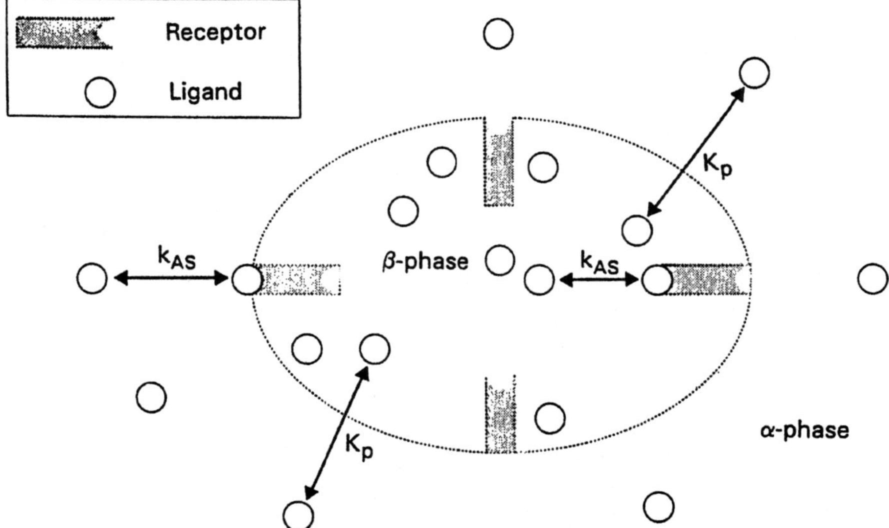

Fig. 7.7. Schematic representation of systems with simultaneous partition and binding of the ligand to illustrate situations in which the orientation of the acceptor site is toward the aqueous (α) phase and toward the lipid (β) phase.

Equations 7.11 and 7.12 are formally identical, and it is therefore impossible to distinguish between the two models depicted in Figure 7.7 purely on the basis of the distribution of ligand between the two phases. In other words, it is not possible to decide whether the parameter derived from the rectangular hyperbolic contribution is the association constant (eq. 7.11) or the product of k_{AS} and the partition coefficient (eq. 7.12).

A convenient procedure for the analysis of systems exhibiting both binding and partitioning requires collection of binding data at concentrations of ligand well in excess of those required for saturation of acceptor sites—a region where the binding process is complete and therefore makes a constant contribution to the binding function. Under those conditions ($k_{AS}C_{\bar{S}}^{\alpha} >> 1$ or $k_{AS}K_pC_{\bar{S}}^{\alpha} >> 1$), eqs. 7.11 and 7.12 both revert to

$$r = K_pC_{\bar{S}}^{\alpha} + p \tag{7.13}$$

Thus, extrapolation of the linear part of the binding curve at high values of $C_{\bar{S}}^{\alpha}$ to the ordinate axis provides the value of p (the number of sites on the acceptor) from the intercept, whereas the partition coefficient (K_p) is obtained from the slope (Fig. 7.8). Once the partition contribution has been defined quantitatively, it can be subtracted from the overall binding curve, whereupon the resultant binding curve corrected for partition may be analyzed according to the standard procedures.

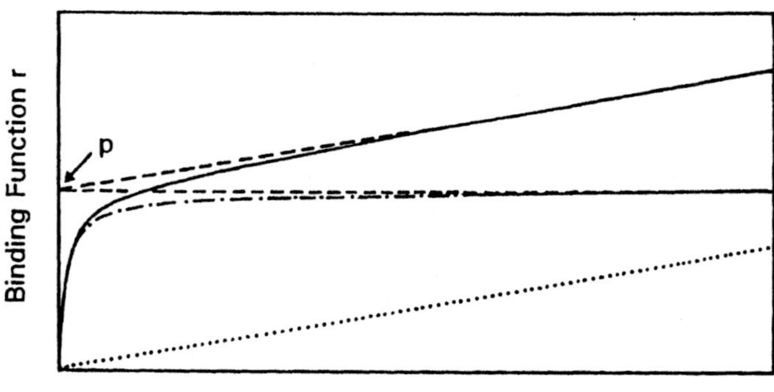

Fig. 7.8. Schematic representation of the effect of ligand partition on the form of a binding curve (solid line). The dashed line indicates the extrapolation of the linear portion to obtain the number of acceptor sites (p) and also to define the plot reflecting partition alone (dotted line). The corrected, rectangular hyperbolic binding curve (dashed, dotted line) is then obtained as the difference between the experimental and partition plots.

7.3.3. Dependence of Binding Function Upon Membrane Concentration

An important consequence of eqs. 7.11 and 7.12 is that the binding function is dependent upon the concentration of membrane, measured in terms of either membrane lipid or membrane protein (Parry *et al.*, 1976). For a reaction mixture with total volume V_t and total substrate concentration $\overline{C}_S$, considerations of mass conservation dictate that

$$V_t\overline{C}_S = V^\alpha C_S^\alpha + V^\beta C_S^\beta \tag{7.14}$$

where V^α and V^β are the respective volumes of the aqueous and membrane phases and where $V^\alpha = (V_t - V^\beta) \approx V_t$. The total and aqueous phase concentrations of free ligand are then related by the expression

$$C_S^\alpha = \overline{C}_S V_t/(V_t + K_p V^\beta) = \overline{C}_S/(1 + K_p \overline{v}_m c_m) \tag{7.15}$$

where c_m is the weight concentration of membrane (expressed, for example, on a protein basis) and $\overline{v}_m$ is the partial specific volume of the membrane expressed on the same basis (Bogoyevitch *et al.*, 1987). An increase in the membrane concentration (*i.e.*, volume of membrane phase at constant total ligand concentration) thus gives rise to a decrease in C_S^α and hence to the magnitude of the binding function (eq. 7.11 or 7.12).

7.2.4. Relevance to Membrane-Bound Enzymes

A number of membrane-bound enzymes act on hydrophobic substrates that can partition into the membrane phase. For example, the microsomal cytochrome

P-450 system is involved in the ω-oxidation of fatty acids, the synthesis of sterols, and the detoxification of a wide range of drugs. We therefore direct attention to the effect of simultaneous binding and partitioning processes on the kinetic properties of such enzymes.

In experimental studies the measured Michaelis constant, $\overline{K}_m$, is determined on the basis that the initial velocity, v, is related to the maximal velocity, v_m, by

$$v = v_m \overline{C}_S/(\overline{K}_m + \overline{C}_S) \tag{7.16}$$

where $\overline{C}_S$ is the concentration of ligand (substrate) defined on the basis of total volume. Because the concentration of substrate greatly exceeds that of enzyme, we may neglect the contribution of enzyme–substrate complex to $\overline{C}_S$, whereupon eq. 7.15 provides the relationship for C_S^α, the aqueous concentration of free ligand. If the active site of a cytochrome P-450 isoenzyme is exposed to the aqueous phase, C_S^α is the concentration of substrate seen by the enzyme; and the expression for initial velocity in terms of the corresponding Michaelis constant, K_m^α, becomes

$$v = v_m \overline{C}_S/[K_m^\alpha(1 + K_p \overline{v}_m c_m) + \overline{C}_S] \tag{7.17}$$

whereupon the measured Michaelis constant ($\overline{K}_m$) and the Michaelis constant describing the interaction in the aqueous phase (K_m^α) are related by

$$\overline{K}_m = K_m^\alpha(1 + K_p \overline{v}_m c_m) \tag{7.18}$$

Similar consideration of the situation in which the enzyme active site is directed toward the lipid phase shows that the counterpart of eq. 7.18 is

$$\overline{K}_m = K_m^\beta(1 + K_p \overline{v}_m c_m)/K_p \tag{7.19}$$

where K_m^β describes the interaction of enzyme with ligand in the membrane phase.

The use of the above expressions is illustrated with results from a kinetic study of the hydroxylation of benzo[α]pyrene by rat liver microsomes (Cumps *et al.*, 1977), which yielded an essentially linear dependence of $\overline{K}_m$ upon microsomal protein concentration (Fig. 7.9a). The ordinate intercept signifies a value of 0.32 μM for K_m^α in the unlikely event that the active site of the microsomal cytochrome P-450 isoenzyme for benzpyrene hydroxylation is directed toward the aqueous phase. Alternatively, the Michaelis constant for reaction with an active site in the microsomal phase is obtained by multiplying this value by K_p (see eqs. 7.18 and 7.19). Combining the microsomal partial specific volume of 0.85 ml/g protein (Bogoyevitch *et al.*, 1987) with the slope of Figure 7.9a gives, via eq. 7.18, a value of 1.3×10^5 for the partition coefficient, K_p. A Michaelis constant (K_m^β) of 42 mM is thus obtained for

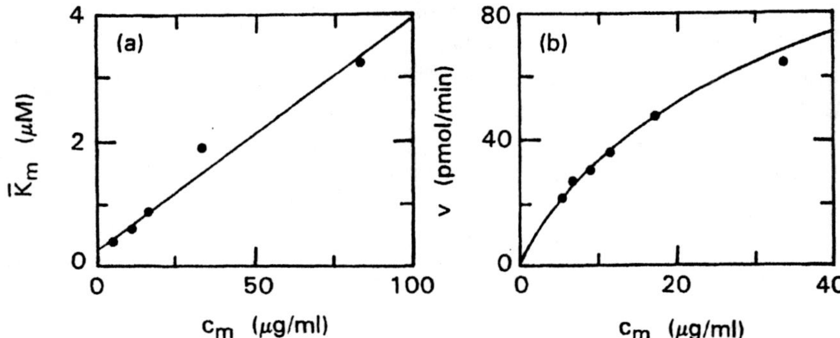

Fig. 7.9. Effect of microsomal protein concentration on the kinetics of benzpyrene hydroxylation by rat liver microsomes, the experimental points for the concentration dependences of (**a**) the measured Michaelis constant ($\overline{K}_m$) and (**b**) the initial velocity in reaction mixtures containing 0.67 μM substrate being inferred from Figures 5A and 4 of Cumps *et al.* (1977), respectively.

interaction of the arylhydrocarbon with a cytochrome P-450 active site located within the membrane phase.

Another consequence of substrate partition is the nonlinear dependence of initial velocity upon enzyme concentration, which can only be varied by changing the membrane concentration. This effect, which is identifiably different from the behavior of a soluble enzyme, is evident in Figure 7.9b, which summarizes the velocity of 3-hydroxybenzpyrene formation in mixtures containing a fixed benzpyrene concentration ($\overline{C}_S$) and a range of microsomal protein concentrations (Cumps *et al.*, 1977). The line in Figure 7.9b is the dependence predicted by eq. 7.17 on the basis of the linear dependence of $\overline{K}_m$ upon $\overline{C}_S$ (Fig. 7.9a) and the value of 6.6 nmol/min/mg protein for the specific activity, v_m/c_m, of the microsomal suspensions (Cumps *et al.*, 1977).

7.4. NONSPECIFIC BINDING

The binding of ligands to membrane-bound acceptors (receptors) forms a fundamental area of study in cell biology, immunology, and endocrinology. Most studies are directed toward the determination of binding affinities and capacities in order to understand how these parameters relate and contribute to control processes *in vivo*. Usually, a radioactively labeled ligand is used so that the concentration of unbound ligand can be monitored in the supernatant after centrifugal separation of the membranes. Besides binding to specific sites on cell surface receptors, most ligands also bind nonspecifically to the membrane surface. The operational criterion used to distinguish the specific from the nonspecific binding is that the specifically bound ligand can be displaced from the membrane by an excess of nonradioactively labeled ligand, whereas the nonspecifically bound ligand is unaffected by this procedure. The nonspecific component thus has the characteristics of partition, and the concentration of nonspecifically bound ligand is linearly related to C_S^α, the free ligand concentration in the aqueous phase. The binding function thus becomes the sum of the partition (nonspecific) and binding (specific) terms (eq. 7.11).

The important point to note at this stage is that the partitioning and binding phenomena respond differently to the addition of an excess of ligand to the system. This difference forms the basis of a useful procedure for distinguishing specific binding of a ligand from nonspecific uptake of ligand by the particulate phase in studies of interactions involving cell surface receptors.

Consideration must be given to the concentration of unlabeled ligand that is required to displace the specifically bound labeled ligand. As a general rule the unlabeled competing ligand should occupy at least 99% of the acceptor sites even at the highest concentration of labeled ligand. In practice, this means that the concentration of unlabeled ligand should be at least two orders of magnitude greater than the dissociation constant ($1/k_{AS}$) of the labeled ligand.

Thus, the nonspecific component of ligand binding may be determined in two ways: (1) by extrapolating the linear part of the binding curve in the region of acceptor-site saturation (Fig. 7.8) and (2) by determining the concentration of ligand bound on addition of an excess of unlabeled ligand. The equivalence of these two approaches is demonstrated in Figure 7.10 for the binding of ^{125}I-labeled insulin to the insulin receptor of Chinese hampster ovary cells. The similarity in slopes of the binding curves under conditions of excess labeled ligand and excess unlabeled ligand is immediately apparent.

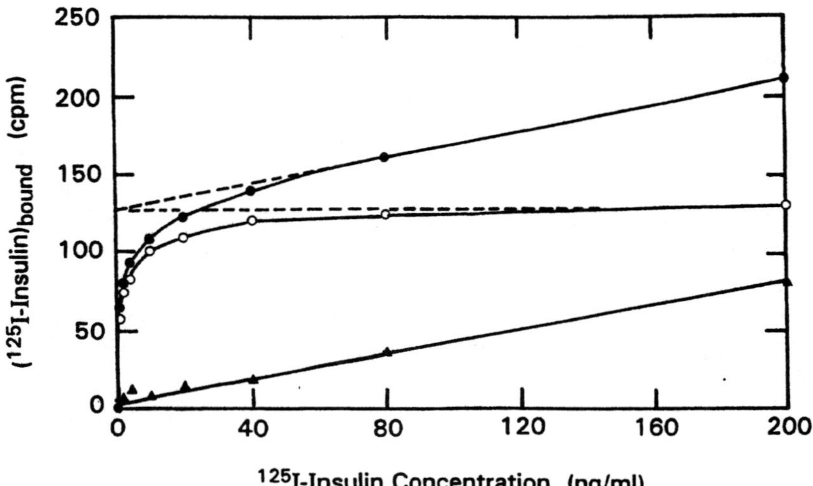

Fig. 7.10. Results of two studies of the binding of [^{125}I] insulin to insulin receptors on Chinese hampster ovary cells (CHO-T cells) to show the equivalence of the two methods for determining nonspecific binding, *viz.*, excess labeled ligand (●), or excess unlabeled ligand (▲) (S. Lucas, and W. H. Sawyer, unpublished results). The difference between these curves (○) describes specific binding.

Because nonspecific binding is nonsaturable and therefore continues to increase with increasing ligand concentration, it is frequently more of a problem when the specific binding is relatively weak, whereupon higher concentrations of ligand are required to saturate specific sites on the acceptor. Difficulties also arise when strong and weak binding sites coexist. The concentration of excess cold ligand may suffice to displace the labeled ligand from the high-affinity but not the low-affinity sites, in which case the contribution of the latter would be classified as nonspecific binding.

In studies of the binding of ligands by whole cells, the physical occlusion of ligand into endocytic or exocytic vesicles may well contribute to nonspecific binding. Moreover, hydrophobic ligands may partition into the plasma membrane, equilibrate with the cytoplasm of the cell, and thereby become associated with subcellular structures. A nonbiological source of ligand uptake symptomatic of nonspecific binding entails its irreversible adsorption to surfaces of assay vessels and filters, a contribution that may be taken into account by performing suitable background control experiments.

Finally, it cannot necessarily be assumed that the extent of nonspecific binding is identical for labeled and unlabeled ligands. The chemical modification of a ligand by attachment of a fluorescent probe will frequently render the ligand more "sticky," thereby increasing its tendency to partition and hence contribute to the nonspecific component of ligand binding. This consideration assumes particular importance when fluorescence parameters of the bound ligand (*e.g.*, anisotropy, emission wavelength, resonance energy transfer) are being used to measure properties of the ligand–receptor complex.

7.5. MOLECULAR CROWDING EFFECTS

Throughout these discussions of the characterization of ligand binding, the law of mass action has been written in terms of concentrations rather than thermodynamic activities of reacting species, a course of action that assumes thermodynamic ideality of the system. Inasmuch as the two concentration scales merge in the limit of infinite dilution (all $C_i \rightarrow 0$), the relatively small concentrations of reactants that are normally used to characterize ligand binding in an *in vitro* study render the assumption an entirely reasonable approximation. However, some consideration needs to be given to the effects of thermodynamic nonideality in the crowded cellular environment.

The thermodynamic activity (a_i) and molar concentration (C_i) of reactant i are linked by the expression

$$a_i = \gamma_i C_i \tag{7.20}$$

where γ_i is the activity coefficient. Under thermodynamically nonideal conditions it is therefore necessary to write the equilibrium constants for ligand binding ($A + S \rightleftarrows AS$) and acceptor self-association ($nM \rightleftarrows P$) as

$$K = a_{AS}/(a_A a_S) = [C_{AS}/(C_A C_S)][\gamma_{AS}/(\gamma_A \gamma_S)] = K'[\gamma_{AS}/(\gamma_A \gamma_S)] \quad (7.21a)$$

$$X = a_P/a_M^n = (C_P C_M^n)(\gamma_P/\gamma_M^n) = X'(\gamma_P/\gamma_M^n) \quad (7.21b)$$

which shows that the effective equilibrium constant defined by the appropriate ratio of species concentrations is obtainable from the true thermodynamic constant by dividing by the corresponding ratio of activity coefficients. The magnitude of the activity coefficient, γ_i, of species i depends upon the composition of the whole solution, the relevant expression being

$$\gamma_i = \exp\left[2B_{i,i}C_i + \sum_j B_{i,j}C_j + \cdots\right] \quad (7.22)$$

where $B_{i,i}$ and $B_{i,j}$ are the respective coefficients that describe the nonideality of a species arising from its own presence and arising from the presence of every other species in the solution. The terms in $B_{i,j}$ are especially relevant to the concentrated cellular environment, where an enzyme can exhibit markedly nonideal behavior despite its presence at very low concentration.

A major contributing factor to the thermodynamic nonideality of macromolecular solutes is that no two molecules can occupy the same space at the same time, whereupon there is a mutually excluded volume within which their two centers of mass cannot coexist. For two uncharged spherical molecules with radii R_i and R_j this excluded volume (Fig. 7.11) is (McMillan and Mayer, 1945)

$$U_{i,j} = (4/3)\pi N(R_i + R_j)^3 \quad (7.23)$$

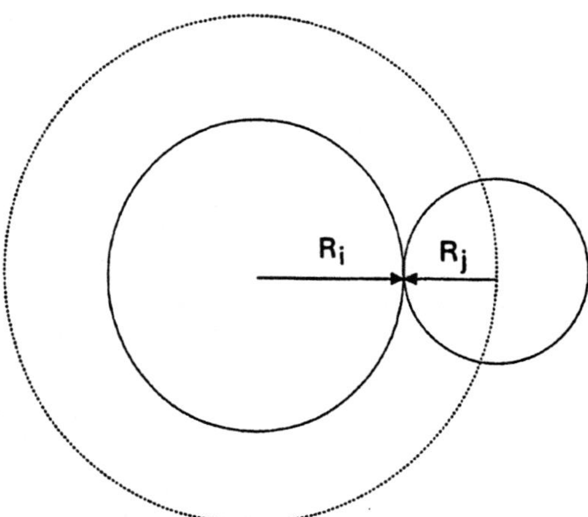

Fig. 7.11. Geometrical illustration of the excluded volume concept that the distance of closest approach for the centers of mass of two spherical molecules is the sum of their radii (eq. 7.23).

in which Avogadro's number (N) is incorporated to express the excluded volume on a molar basis (rather than a molecular basis) to conform with the concentration scale used to define the equilibrium constant. To a first approximation these excluded volumes may be substituted for the virial coefficients in eq. 7.22, which then become (on noting that $U_{i,i} = 2B_{i,i}$)

$$\gamma_i = \exp\left[U_{i,i}C_i + \sum_j U_{i,j}C_j + \cdots\right] \tag{7.24}$$

In a very dilute solution the excluded volume contribution represents a very small fraction of the total volume, and hence little error is introduced by considering that the molecules are distributed throughout the total volume. However, in an extremely concentrated macromolecular environment (such as that prevailing in the red blood cell), the excluded volume contribution becomes a significant proportion of the total volume, which is therefore an overestimate of the effective volume in which the macromolecules are distributed. The molar concentration (C_i) is thus an underestimate of the thermodynamic concentration (a_i); or, in other words, the activity coefficient is greater than unity.

Let us consider the commonly encountered situation in which a protein and its biospecific ligand are present at relatively low concentrations but reacting in a highly concentrated macromolecular environment, which we shall (for simplicity) regard as comprising a single space-filling macromolecule, F, present at concentration C_F. For a small ligand it is more appropriate to consider the magnitude of a_S as a separate issue (Ford *et al.*, 1984; Bergman and Winzor, 1989a; Bergman *et al.*, 1989) because other factors such as electrostatic behavior may well negate the estimate of γ_S deduced on an excluded volume basis. Indeed, to a first approximation the activity coefficient of the ligand may often be taken as unity, whereupon the relationship between the thermodynamic and effective equilibrium constants (K and K', respectively) becomes

$$K' = K \exp\left[(U_{A,F} - U_{AS,F})C_F + \cdots\right] \tag{7.25}$$

Provided that the attachment of ligand to the acceptor results in no marked conformational change in the latter, there should be essentially no difference between R_{AS} and R_A and hence no difference between $U_{A,F}$ and $U_{AS,F}$. The equilibrium constant obtained under conditions approaching thermodynamic ideality (K) should thus also approximate the effective equilibrium constant (K') required for predicting the equilibrium distribution of A, S, and AS in the biological context.

If, however, a ligand-induced change in acceptor conformation accompanies complex formation, the effective binding constant (K') is either enhanced or diminished, depending on the relative sizes of A and AS. In the case of yeast hexokinase, for example, X-ray crystallographic studies (Bennett and Steitz, 1980) have shown that complex formation with glucose is accompanied by a

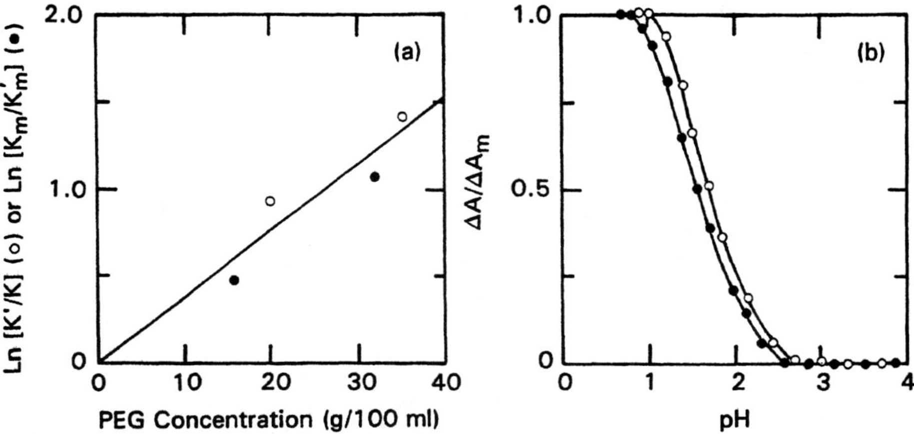

Fig. 7.12. Displacement of ligand-induced isomerization equilibria by an inert solute (*F*). (**a**) Effect of poly(ethylene glycol) concentration on the relative Michaelis and binding constants for the interaction of glucose with yeast hexokinase, the results being inferred from Figure 2 of Rand *et al.* (1993). (**b**) Effect of sucrose on the pH dependence of the unfolding of ribonuclease, as detected by the fractional change in absorbance at 287 nm: ○, the titration in the absence of inert solute; ●, the corresponding titration in the presence of sucrose (0.5 M). [Adapted from Shearwin and Winzor (1990).]

decrease in volume ($R_{AS} < R_A$)—a factor also evident (Winzor and Wills, 1995) from the activating effect of poly(ethylene glycol) on both the binding and catabolism of glucose (Fig. 7.12a). Other enzyme systems for which molecular crowding effects have been used to detect substrate-induced conformational changes include lactate dehydrogenase (Nichol *et al.*, 1983; Bergman and Winzor, 1989a), urease (Nichol *et al.*, 1985) and invertase (Shearwin and Winzor, 1988a). For the last-named enzyme sucrose is both the substrate and the inhibitory space-filling solute.

The molecular crowding effect of a high concentration of an inert solute may also be used (1) to displace pre-existing isomerizations in favor of the smaller isomer, such displacement having been used to demonstrate the pre-existence of the isomerization equilibrium as the source of the allostery exhibited by pyruvate kinase (Harris and Winzor, 1988b); and (2) to establish the equilibrium nature of the acid-induced unfolding of ribonuclease (Fig. 7.12b). Whereas the molecular crowding effect of an inert solute is to displace an isomerization toward the smaller isomer, the corresponding effect on a self-association equilibrium is displacement in favor of polymer (Nichol *et al.*, 1981a; Minton and Wilf, 1981; Shearwin and Winzor, 1988b; Cann *et al.*, 1994). Such effects of high concentrations of ribonuclease (*F*) on the self-association of glyceraldehyde-3-phosphate dehydrogenase (*A*) are shown in Figure 7.13.

As well as indicating the potential of thermodynamic nonideality as a probe of conformational changes in proteins and enzymes, these illustrative examples

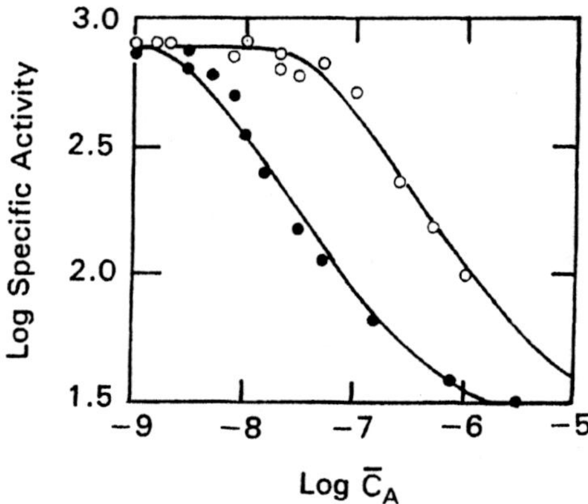

Fig. 7.13. Effect of ribonuclease (*F*) on the self-association of glyceraldehyde-3-phosphate dehydrogenase, as monitored by the dependence of specific activity upon enzyme concentration: ○, results obtained in the absence of inert solute; ●, results obtained in the presence of ribonuclease (13.2 mM). [Adapted from Minton and Wilf (1981).]

also signify the need to consider the likely consequences of molecular crowding in cellular situations (Minton, 1983; Bergman and Winzor, 1989b). We therefore attempt to predict the effect of 5 mM hemoglobin (roughly its concentration in the erythrocyte) on the exclusive binding of substrate to the *R* state of an enzyme undergoing pre-existing isomerization (Fig. 6.4 with $q = 0$). From eqs. 6.10 and 7.2b, it follows that

$$v/v_m = r/p = \frac{k_{RS}C_S(1 + k_{RS}C_S)^{p-1}}{(1 + k_{RS}C_S)^p + X'} \tag{7.26a}$$

$$X' = X \exp [(U_{R,F} - U_{T,F})C_F] \tag{7.26b}$$

provided that ligand binding to the *R* isomer has a negligible effect on its effective radius. Because the binding constant (k_{RS}) and ligand concentration (C_S) invariably appear as the product in eq. 7.26a, $k_{RS}C_S$ is treated as a single variable in the simulations.

Figure 7.14 summarizes the results of numerical simulations based on parameters for pyruvate kinase, a tetrameric enzyme ($p = 4$) with effective isomeric radii of 5.0 and 5.2 nm, to demonstrate the potential space-filling effect of this high concentration of hemoglobin ($R_F = 3.1$ nm) in situations where *R* is the smaller and larger isomer: *X* has been assigned a magnitude of 1,000, a realistic value for pyruvate kinase in the presence of its allosteric inhibitor, phenylalanine. Comparisons of these sigmoidal curves with that simulated for enzyme in the absence of the space-filling protein point to the potential regulatory role of molecular crowding as either an allosteric activator or inhibitor, depending on whether the active form of the enzyme is the smaller or larger isomer. Such displacement of a sigmoidal response by molecular

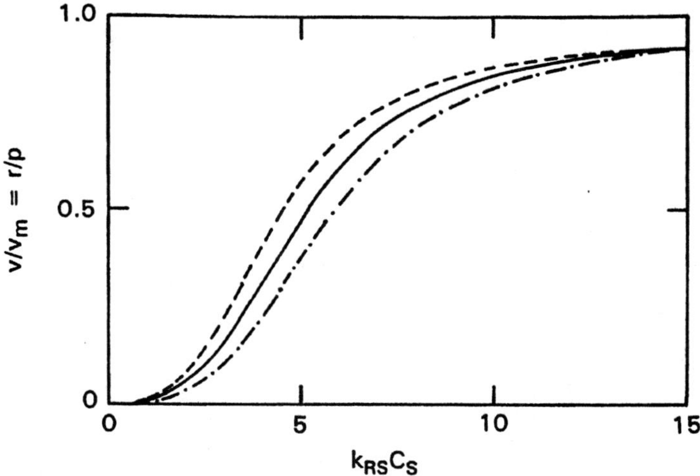

Fig. 7.14. Numerically simulated illustration of the potential of molecular crowding to act as an allosteric effector in the concentrated cellular environment. Solid line, binding curve calculated on the basis of eq. 7.26, with $p = 4$ and $X = 1{,}000$ in the absence of inert solute ($C_F = 0$): dashed and dashed-dotted lines, corresponding curves calculated for the same system in the presence of 5 mM hemoglobin in situations where the active isomer (R) is the smaller and larger acceptor state, respectively.

crowding may either supplement or oppose the allosteric effects arising from interactions of specific effectors with the enzyme (see Chapter 6). However, whereas the specific binding of an allosteric effector to a reversibly isomerizing enzyme has the potential to displace the isomerization equilibrium toward either the R or the T state, the effect of molecular crowding is confined to displacement of the equilibrium toward the smaller isomer. This simple example serves to draw attention to the possibility that the catalytic performance of allosteric enzymes *in vivo* may be modified by factors no more specific than the ability of unrelated solutes to occupy space in the highly concentrated cellular environment (Minton, 1983; Bergman and Winzor, 1989b).

8

STUDIES OF DNA–LIGAND INTERACTIONS

The interaction of ligands with DNA is of fundamental importance in modern molecular biology. Repressors, activators, transcription factors, and enzymes all contribute to control DNA synthesis, repair, degradation, transcription, and gene expression. The binding of inorganic ions to DNA helps to stabilize and maintain the shape of the double helix, while the binding of drugs to DNA forms the basis of many cancer chemotherapies. In preceding chapters the attention has centered on the binding of ligands to globular proteins, in which a combination of secondary and tertiary structures serves to create the specific ligand-binding site(s) involving well-separated residues of the polypeptide chain. Because DNA is a linear acceptor comprising very similar residues (nucleotides), there is a definite potential for random electrostatic interaction with, and also random intercalation into, the polynucleotide duplex. Such random binding of ligand is described as *nonspecific* because of the absence of any unique nucleotide sequence to which the ligand binds; but it must be realized that such use of the term should not be confused with its other meaning in the context of unsaturable uptake of ligand (Sections 7.3 and 7.4). In addition to this nonspecific binding, there can also be interactions of proteins with unique nucleotide sequences, the *lac* operon and *lac* repressor sites being notable examples. However, even these interactions are facilitated by nonspecific interactions of the same ligand with the polynucleotide chain, which allows restricted diffusion of the ligand along the polymer chain towards the specific site (Riggs *et al.*, 1970a; Richter and Eigen, 1974; Schranner and Richter, 1978). Because nonspecific (random) attachment of the ligand to the polynucleotide figures so prominently in DNA–protein interactions, a general discussion of the phenomenon is in order.

8.1 PROTEIN–POLYMER INTERACTIONS

In interactions between proteins and linear polymeric acceptors such as polynucleotides (Latt and Sober, 1967; McGhee and von Hippel, 1974) or polysaccharides (Olson *et al.*, 1991), a complication arises whenever the site for ligand binding comprises a sequence of two or more consecutive residues on the polymer chain. Because of the similarity of the repeating units of the linear polymer chain, there is no specific site for binding. Any residue sequence of the appropriate length is a potential site for ligand attachment. During the early stages of nonspecific binding the random attachment of ligand molecules leads to the existence on the acceptor chain of spaces between bound ligands that contain too few residues to accommodate an additional ligand molecule, despite the fact that there would be sufficient unoccupied residues for further ligand attachment if the bound molecules were aligned optimally. This feature is illustrated in Figure 8.1 for the binding of a ligand to three-residue sequences of a 12-unit lattice. Clearly, the maximum number of ligands that can attach to the acceptor chain is four (top situation); but, if random binding leads to the distribution of the first three ligands shown in the center situation, this chain is effectively saturated until the combined processes of dissociation and association result in the required optimal alignment of bound ligand molecules. The bottom situation in Figure 8.1 represents an even more acute parking problem, whereby effective saturation of the lattice is achieved after the attachment of only two ligand molecules.

From the thermodynamic viewpoint the latter two situations are not viable options because they merely represent transient (nonequilibrium) states on the route to chain saturation. The maximum number of ligands that can be attached to the acceptor lattice (p) is unequivocally four. However, although the time course of the elimination of the parking problem is irrelevant to the thermodynamics of nonspecific ligand binding, it is a matter of vital concern to an experimenter, who must wait patiently for the kinetics of this bound-ligand

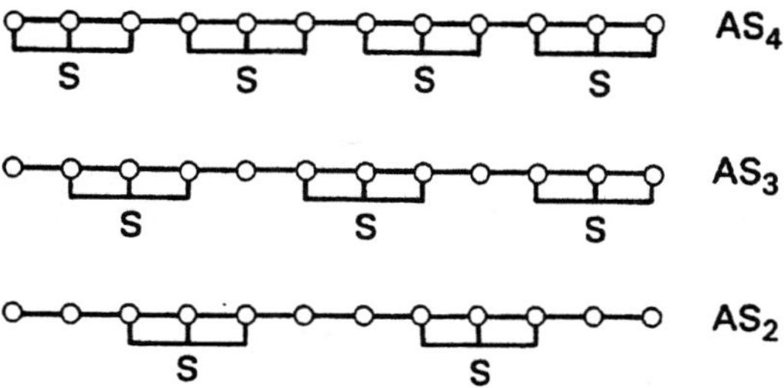

Fig. 8.1. Schematic representation of three "saturated" complexes formed during the early stages of random attachment of a ligand (S) to 3-residue sequences on a 12-residue acceptor lattice (A). The lower two situations are transient states that are eliminated in the attainment of thermodynamic equilibrium.

rearrangement to pursue its course before taking the equilibrium measurement that is required for substitution into the appropriate thermodynamic expression. We therefore first consider this question of the kinetics of the approach to thermodynamic equilibrium.

8.1.1. Kinetics of Nonspecific Binding

Even though the potential for suboptimal use of polymer residues is maximal under conditions conducive to saturation of the acceptor chain with ligand, thermodynamic considerations dictate that the binding function approach the value commensurate with optimal alignment of ligand molecules throughout the length of the acceptor lattice. Considerable time may therefore be required for the attainment of thermodynamic equilibrium. In spectral studies of non-specific binding it is thus important to ascertain whether a seemingly stable binding response is the equilibrium value or merely a pseudo-equilibrium response prior to a much slower phase of thermodynamic equilibrium attainment reflecting the kinetics of elimination of the parking problem. The existence of this slow phase, first reported by Epstein (1979), is evident in Figure 8.2, which summarizes simulated time courses for equilibrium attainment for non-specific binding of a ligand to three-residue segments of a six-residue lattice governed by an association rate constant (k_f) of 10^3 $M^{-1}s^{-1}$ and the indicated values (s^{-1}) of the dissociation rate constant (k_r).

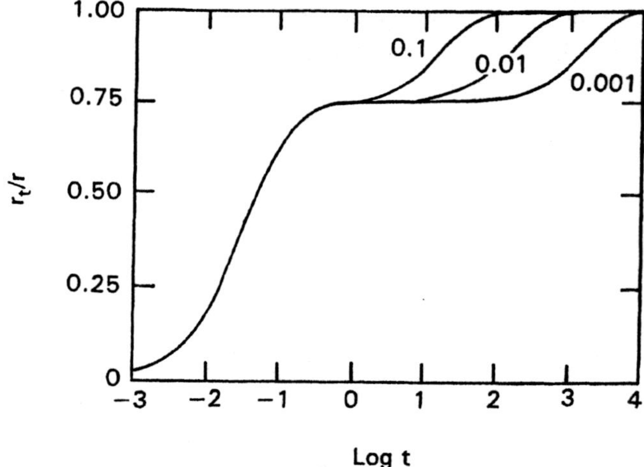

Fig. 8.2. Simulated time dependence of the approach to equilibrium for nonspecific binding of a ligand to three-residue sequences of a six-residue acceptor lattice. Simulations are based on total acceptor and ligand concentrations ($\overline{C}_A$, $\overline{C}_S$) of 1 μM and 100 μM, respectively; an association rate constant (k_f) of 1,000 $M^{-1}s^{-1}$; and the indicated magnitudes (s^{-1}) for the dissociation rate constant, k_r (P. D. Munro and D. J. Winzor, unpublished data).

It can certainly be argued that simulations such as those presented in Figure 8.2 refer to situations in which the association rate constant is several orders of magnitude below the level of diffusion control and that equilibrium in an experimental system with an intrinsic binding constant ($k_{AS} = k_f/k_r$) of 10^5 M^{-1} may well be attained much more rapidly than is inferred from the simulated progress curve ($k_r = 0.01\ s^{-1}$), which indicates a time requirement of about 10 minutes (1,000 seconds). However, even if k_a and k_r were both increased 10-fold (or even 100-fold), the necessity to monitor the binding response for at least 100 seconds (or 10 seconds) would probably not be met if the reaction were being monitored by (say) conventional temperature-jump measurements.

The consequences of the kinetics of eliminating the parking problem can certainly be minimized by restricting measurements of ligand binding to low levels of acceptor saturation. The simulations in Figure 8.2 refer to mixtures for which equilibrium reflects 85% saturation of the acceptor lattice. Unfortunately, such action is not the solution to an experimenter's dilemma, unless the length of the residue sequence for ligand binding (m) can be ascertained by other means. A lack of binding data in the high C_S region of a binding curve clearly precludes unequivocal experimental delineation of the binding capacity of the acceptor lattice. This capacity, $p = N/m$ for an N-residue lattice, is usually the key indicator of the length of the residue sequence (m) occupied by a ligand molecule and hence the determinant of the binding equation to be used to analyze the experimental results. In that regard, the problem of data extrapolation to obtain the acceptor capacity (p) is compounded by the negatively cooperative form of binding curves reflecting nonspecific interaction of a ligand with a polymer lattice—the subject of the next section.

8.1.2. Thermodynamics of Nonspecific Binding

For noncooperative binding of a ligand to a sequence of m lattice residues, the general form of the binding equation is (Epstein, 1978)

$$r = \left[\sum_{1}^{p} i\ {}^{h}C_i k_{AS}^{i} C_S^{i}\right] \Big/ \left[1 + \sum_{1}^{p} {}^{h}C_i k_{AS}^{i} C_S^{i}\right] \tag{8.1}$$

where p is the integer value of N/m, and ${}^{h}C_i = h!/[i!(h - i)!]$ is the number of possible arrangements of the h items (bound ligands and unoccupied lattice residues) available for distribution on the lattice. In the attachment of the ith ligand molecule, the number of items to be arranged is $h = (N - mi + i)$, whereupon (Latt and Sober, 1967)

$${}^{h}C_i = (N - mi + i)!/[i!(N - mi)!] \tag{8.2}$$

The use of eqs. 8.1 and 8.2 is illustrated in Figure 8.3 for the interaction of a ligand with three-unit sequences of a 13-unit acceptor lattice, a situation rele-

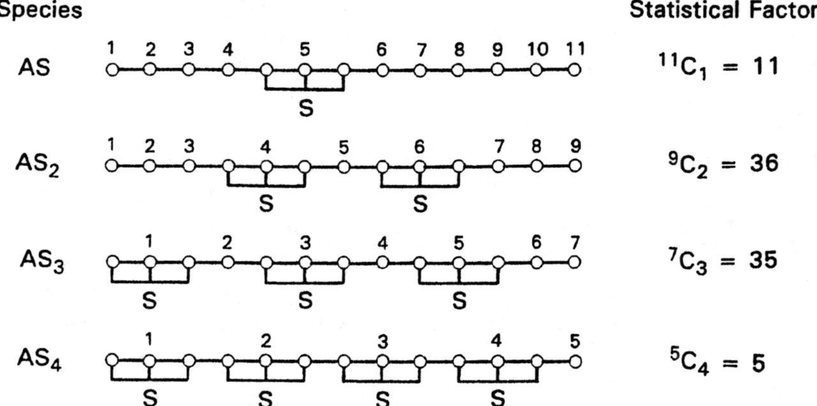

Fig. 8.3. Schematic illustration of the number of items to be arranged, and hence the statistical factors to be used, when a ligand binds nonspecifically to three-residue sequences of a linear acceptor lattice comprising 13 residues.

vant to the nonspecific binding of thrombin to a fractionated heparin preparation (Olson *et al.*, 1991). On the basis of Figure 8.3 the expression for the binding function r (eq. 8.1) becomes

$$r = \frac{11k_{AS}C_S + 72k_{AS}^2C_S^2 + 105k_{AS}^3C_S^3 + 20k_{AS}^4C_S^4}{1 + 11k_{AS}C_S + 36k_{AS}^2C_S^2 + 35k_{AS}^3C_S^3 + 5k_{AS}^4C_S^4} \tag{8.3}$$

Analysis of spectrofluorometric titration data for the interaction of thrombin with a heparin fraction ($N = 13$ disaccharides) in such terms is illustrated in Figure 8.4, where the solid line is the best-fit theoretical relationship obtained by nonlinear regression analysis of the results in terms of eq. 8.3 and the resulting value of 1.0 ($\pm$0.1) $\times$ 10^5 M^{-1} for k_{AS} (Olson *et al.*, 1991).

The first point to note in relation to Figure 8.4 is the negatively cooperative form of a Scatchard plot reflecting nonspecific binding of a ligand to a polymer lattice. As mentioned above, this feature of the binding curve renders prohibitively difficult the task of employing the binding data to define p (the abscissa intercept of the Scatchard plot) because of the lack of experimental results in the crucial region for reliable extrapolation of the data. Knowledge of that abscissa intercept is critical because it determines the magnitudes of the ^hC_i terms (eq. 8.2) and hence the magnitudes of the polynomial coefficients in the binding equation. To overcome this dilemma, Olson *et al.* (1991) deduced the value of m (and hence p) by extrapolating the chainlength dependence of K_{AS} (the stoichiometric binding constant) for the interaction of thrombin with short heparin chains ($2 < N < 6$) to zero binding strength to obtain the segment length ($N = 3$ disaccharides) below which interaction effectively ceased.

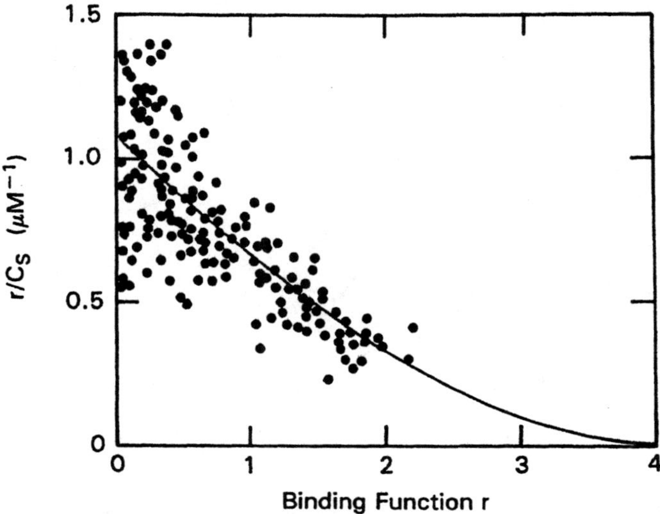

Fig. 8.4. Scatchard plot reflecting the nonspecific binding of thrombin to a fractionated heparin preparation. The line denotes the best-fit relationship obtained by nonlinear regression analysis of the results in terms of eq. 8.3. [Adapted from Olson *et al.* (1991).]

8.2. NONSPECIFIC DNA-LIGAND INTERACTIONS

The above combinatorial approach to the derivation of binding equations become very cumbersome as the length of the acceptor lattice increases and is therefore not well-suited for application to studies with DNA as the linear polymer. Indeed, the necessity to specify the value of p ($= N/m$) is also a decided disadvantage of the combinatorial approach, because the number of nucleotides comprising a DNA chain is (1) usually unknown and (2) almost certainly variable because of heterogeneity of the DNA with respect to chain-length. These difficulties have been overcome (McGhee and von Hippel, 1974) by switching to a probability approach, which avoids the necessity to define p (or N) and thus encompasses any variation in polynucleotide chainlength. Its only disadvantage is an implicit assumption that the chainlength is infinite; but, as noted by Epstein (1978), this approximation is unlikely to introduce serious error even for relatively short lattices, provided that binding is not highly cooperative.

For the random binding of a ligand to m-residue sequences of an infinite acceptor lattice, McGhee and von Hippel (1974) have derived the expression

$$r/C_S = k_{AS}(1 - mr)\{(1 - mr)/[1 - (m - 1)r]\}^{m-1} \qquad (8.4)$$

which has the advantage of being a closed solution (*cf.* eq. 8.3), and which, of necessity, reverts to a Scatchard-type relationship when $m = 1$ because all

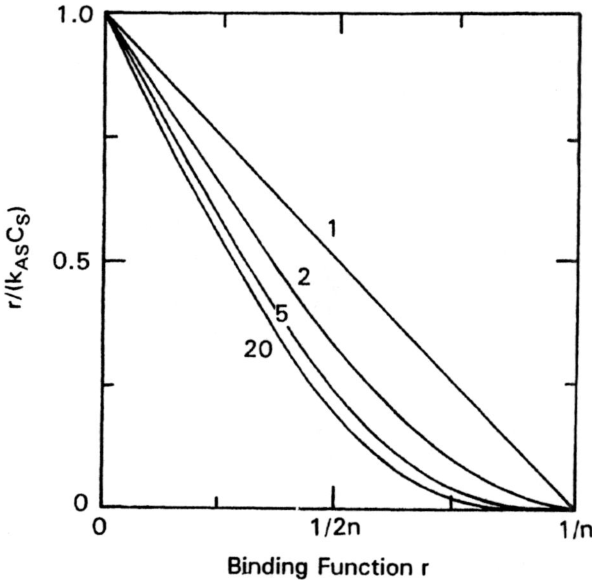

Fig. 8.5. Simulated Scatchard plots for the noncooperative nonspecific binding of ligands to *m*-residue sequences on a polymer of infinite chainlength. The total acceptor concentration used to define the binding function is expressed in terms of nucleotide content, and the number adjacent to each curve denotes the value of *m* that has been used to calculate the dependence via eq. 8.4. [Adapted from McGhee and von Hippel (1974).]

nucleotide residues then become equivalent and independent sites. However, the ordinate intercept of a plot of r/C_S *versus* r for data conforming with eq. 8.4 yields the intrinsic association constant for nonspecific binding, whereas the corresponding parameter is pk_{AS} for equivalent and independent binding to specific sites. These features are evident in Figure 8.5, which shows that the negatively cooperative form of the Scatchard plot for nonspecific binding to a lattice of finite chainlength (Fig. 8.4) also extends to situations in which the acceptor is an infinite lattice.

8.2.1. Allowance for Cooperativity

The probability treatment is readily extended to account for situations in which the occupation of an *m*-residue sequence on the acceptor lattice affects the binding of ligand to adjacent sequences. As in Chapter 6, such binding may be positively or negatively cooperative. When the ligand is a protein, the cooperativity can reflect either protein–protein interactions between adjacent ligands on the linear lattice or ligand-induced distortion of the DNA structure in the immediate vicinity of a bound protein molecule. Positive cooperativity leads to clustering of ligands on the DNA lattice whenever the probability of additional ligand attachment to an adjacent sequence outweighs the probability of random attachment because of the consequent greater affinity of an adjacent sequence for ligand.

Cooperativity is introduced into the model by designating the modified affinity of a sequence adjacent to a bound ligand molecule as ωk_{AS} and that for a flanked m-residue sequence by $\omega^2 k_{AS}$. Values of ω greater and less than unity are the respective symptoms of positive and negative cooperativity. Such extension of the probability treatment of nonspecific binding leads to the expression (McGhee and von Hippel, 1974)

$$\frac{r}{C_S} = \frac{k_{AS}(1 - mr)[(2\omega - 1)(1 - mr) + r - R]^{m-1}[1 - (m + 1)r + R]^2}{[2(\omega - 1)(1 - mr)]^{m-1}[2(1 - mr)]^2} \tag{8.5a}$$

$$R = \{[1 - (m + 1)r]^2 + 4\omega r(1 - mr)\}^{1/2} \tag{8.5b}$$

Analysis of data in instances where m is of unknown magnitude can be very difficult because any decrease in the extent of curvature of the Scatchard plot as the result of positive cooperativity may also be rationalized by decreasing the length of the putative nucleotide sequence to which ligand binds noncooperatively (Fig. 8.5). Similarly, the operation of negative cooperativity can be accommodated by increasing the length of the nucleotide sequence for noncooperative binding. Unless the value of m is available from other sources, the analysis frequently only serves to define the ratio m/ω (Epstein, 1978). For systems with very high positive cooperativity ($\omega > 10$m) the binding curve assumes a distinctly sigmoidal form; and hence the existence of positive cooperativity ceases to be a matter of conjecture, provided that the ligand possesses only one binding site (see Section 7.1.2). However, the dilemma of employing equilibrium titration data alone to evaluate m, ω, and k_{AS} remains.

The problem has been addressed in depth by Kowalczykowski *et al.* (1986), who recommend that an additional titration be performed under essentially stoichiometric conditions ($k_{AS} \rightarrow \infty$) to allow the separate identification of m. The same recommendation was made in the earlier discussion of spectral titration procedures for characterizing specific binding (Chapter 3). For DNA–protein interactions the second titration is frequently performed at lower ionic strength to effect the required enhancement of k_{AS} (Record *et al.*, 1981). However, the experimenter should bear in mind the fact that the much larger binding constant greatly increases the likelihood of an extremely slow rate of complex dissociation (k_r). Extreme precautions are therefore required to verify that the endpoint of the titration does reflect thermodynamic equilibrium, and not merely the irreversible (pseudo-equilibrium) endpoint prior to the elimination of the parking problem (Section 8.1.1).

Once the length of the nucleotide sequence for ligand binding (m) has been ascertained, advantage is taken of the finding (McGhee and von Hippel, 1974) that for systems with $\omega > 10m$ the free ligand concentration associated with 50% lattice saturation defines the reciprocal of the product of the binding constant and the cooperativity factor, *i.e.*, $(C_S)_{0.5} = 1/(\omega k_{AS})$. In the event that the abscissa of the binding curve is plotted in terms of total rather than free ligand concentration, the corresponding expression becomes

$$(1/\omega k_{AS}) = (\overline{C}_S)_{0.5} - \overline{C}_A(N/2m) \tag{8.6}$$

if the concentration of the acceptor lattice is defined in molar terms. In that connection it should be noted that $N\overline{C}_A$ defines the lattice concentration in terms of nucleotide content, the conventional means of expressing the composition of DNA or polynucleotide solutions. Identification of the product ωk_{AS} on the basis of eq. 8.6 then allows the separate magnitudes of ω and k_{AS} to be assessed by comparing the experimental equilibrium titration curve with those generated for a range of ω and corresponding k_{AS} values.

This approach to the problem of characterizing positively cooperative binding of a ligand to a polynucleotide lattice is illustrated in Figure 8.6 for the interaction between poly(riboethenoadenylic acid) and gene 32 protein (Kowalczykowski *et al.*, 1981, 1986; Newport *et al.*, 1981). On the basis that the stoichiometric titration (Fig. 8.6a) signifies a binding sequence of seven residues ($m = 7$), the midpoint, $(\overline{C}_S)_{0.5}$, of the equilibrium titration curve (Fig. 8.6b) is combined with the polynucleotide concentration ($N\overline{C}_A$) of 1 μM to obtain an estimate of 7×10^6 M^{-1} for ωk_{AS} (eq. 8.6). From the simulated curves for a range of ω values, it was concluded that the combination of ω = 1,500 and k_{AS} = 4,700 M^{-1} provided the best fit of the results (Kowalczykowski *et al.*, 1986).

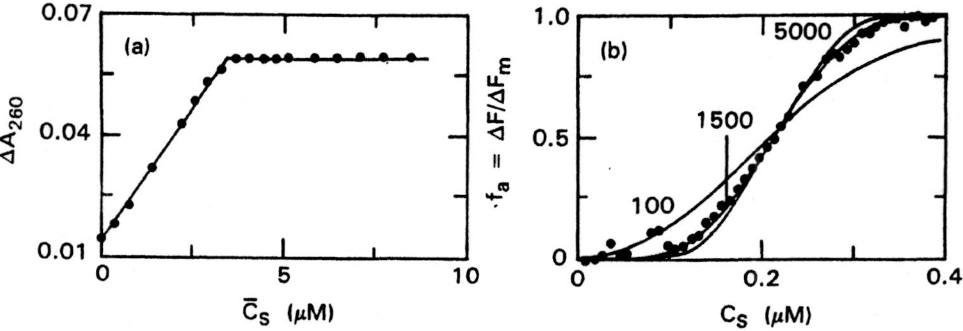

Fig. 8.6. Illustration of the approach for characterizing cooperative nonspecific binding of a protein ligand to a polynucleotide. **(a)** Titration of poly(riboethenoadenylic acid) with gene 32 protein under stoichiometric conditions to obtain the size (m) of the protein-binding site on the acceptor lattice. [Adapted from Newport *et al.*, (1981)]. **(b)** Evaluation of the binding constant (k_{AS}) and cooperativity factor (ω) from a corresponding equilibrium titration, monitored spectrofluorimetrically. Solid lines have been generated on the basis of a magnitude of 7×10^6 M^{-1} for ωk_{AS} and the indicated values of ω. [Adapted from Kowalczykowski *et al.* (1981).]

8.2.2. The Prevalence and Importance of DNA–Protein Interactions

Nonspecific binding may be of importance *in vivo* since it provides a mechanism whereby a specific DNA-binding protein can search for a unique sequence of base pairs that makes up its binding site. The on-rate for the binding of a protein to its specific site on a DNA lattice can thus be much faster than would be expected theoretically (Richter and Eigen, 1974). It therefore seems that the reduction in dimensionality provided by nonspecific binding allows the protein to move laterally along the lattice until the specific site is found. Alternatively, dissociation from a nonspecific site can lead to reassociation on an adjacent domain of the random coil or tightly packed polymer (Berg *et al.*, 1982). In the case of oligomeric repressors, the protein can cross-link two distant sites on the DNA. If the affinity of each unit in the repressor for DNA is equal, the protein has a 50% chance of being transferred between domains. An important way of distinguishing between specific and nonspecific binding is on the basis of salt sensitivity. For example, the nonspecific complex between *cro* repressor and DNA is dissociated completely by 0.4–0.5 M NaCl, whereas 1.6 M NaCl is required to dissociate the *cro* repressor from its specific operator site (Boschelli *et al.*, 1982). Molecular exclusion effects (see Section 7.5) can also be important in increasing the association constant of protein–DNA interactions *in vivo* (Jarvis *et al.*, 1990).

Numerous proteins can bind to DNA regardless of the base sequence. Among these are the histones, the *Escherichia coli* single chain binding protein (Lohman *et al.*, 1986; Bujalowski and Lohman, 1987), the bacteriophage gene 32 protein (Kowalczykowski *et al.*, 1981), and the bacterophage gene 5 protein (Porschke and Rauh, 1983). Proteins that recognize specific base sequences on DNA are involved in the control of gene expression and have therefore received wide attention. A number of conformational motifs in sequence-specific DNA-binding proteins have been identified: zinc fingers, leucine zippers, and helix-turn-helix motifs, to name three. Methods for identifying and quantifying DNA–protein interactions are therefore of great importance. Thermodynamic characterization of these interactions has the potential to provide information about the types of bonds providing the interaction energy and about the effects of mutations, mismatches, insertions, and deletions in the DNA sequence. Because the protein ligand is itself a macromolecule, many of the methods described earlier are inapplicable. Other experimental procedures have therefore evolved to overcome this problem.

8.3. THE SPECTROSCOPIC STUDY OF DNA–PROTEIN INTERACTIONS

Absorption spectroscopy has not been used extensively to study DNA–protein interactions due to the spectral overlap of DNA and protein. However, in cases where the absorption of either component is minimal, difference spectroscopic techniques can be used to determine binding parameters and to provide some information about the microenvironment of the chromophores. For nonspecific DNA-binding interactions the absorption of all bases may be perturbed by binding along the length of the polymer. For specific interactions the perturbations in a single contact sequence may be difficult to detect because of the large background absorption of the DNA. A notable exception is the hyperchromicity that results from the unwinding of 5–15 base pairs on DNA due to the binding of RNA polymerase. Even here it is desirable to have about one polymerase molecule bound per 100 base pairs (Shimer *et al.*, 1988). A possibly more rewarding alternative is to use difference spectroscopy to detect perturbation of the protein structure. For example, it has been shown that at least one of the tyrosine residues in the gene 5 protein of bacteriophage M13 is in a less polar environment after complex formation with single stranded DNA (van Amerongen *et al.*, 1990). Such spectral changes can, in theory, be used to quantify binding, as exemplified by the application of difference spectroscopy to characterize the interactions of gene 32 protein with poly-(riboethenoadenylic acid) (Kowalczykowski *et al.*, 1981).

Fluorescence spectroscopy offers greater flexibility in experimental design and provides sufficient sensitivity to be compatible with studies in the concentration region approaching the dissociation constant ($1/k_{AS}$) of many DNA–protein interactions. When an aromatic amino acid is located within the DNA-binding domain of the protein, its quantum yield or emission maximum will frequently be altered on binding to DNA. These parameters can therefore be used to quantify the binding reaction. To clarify the interpretation, it is sometimes possible to engineer a protein genetically so that is possesses a single tryptophan residue within the DNA-binding motif (Hard *et al.*, 1989; He and Mathews, 1990; Gardner and Mathews, 1990). DNA itself is nonfluorescent at room temperature; but chemistries are now available that permit fluorescent probes to be conjugated at specific positions in the base sequence, thereby enabling "fluorescent footprinting" of the DNA (Guest *et al.*, 1991; Jameson and Sawyer, 1995). Judicious choice of the labeling position should allow changes in spectral characteristics to be used to monitor protein binding without interference to the binding equilibrium itself.

Three types of binding titration are commonly used in these situations. A direct method involves the addition of protein to a fixed amount of DNA. If the intrinsic fluorescence is quenched on binding, the difference between the signal in the presence and absence of protein may be employed as a measure of the degree of saturation of binding sites (Alma *et al.*, 1983). The second method involves the addition to DNA to a fixed amount of protein, whereupon spectroscopic changes in the ligand (protein) may be used to monitor the association. The theory pertaining to both of these types of titration has been presented by Kowalczykowski *et al.* (1986). The third method involves isoparametric analysis (see Section 3.5) of a set of binding curves obtained by the addition of DNA to several fixed concentrations of protein (Lohman and Mascotti, 1992).

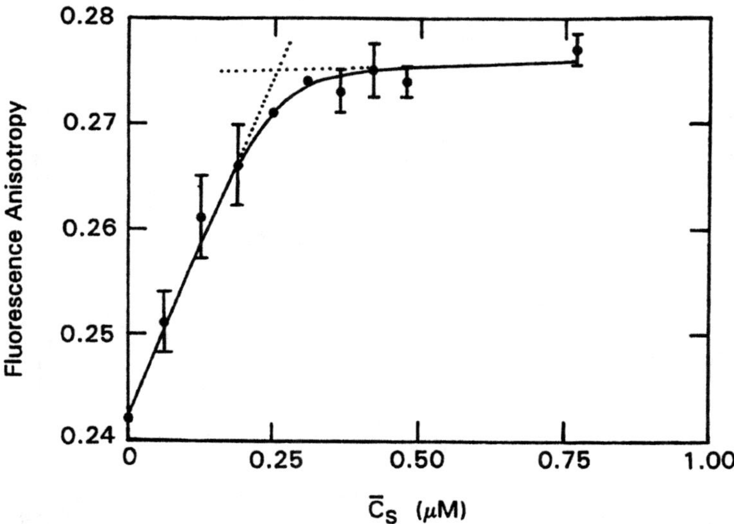

Fig. 8.7. The use of fluorescence anisotropy to measure the interaction of the TyrR repressor protein of *Escherichia coli* with a coumarin-labeled 42-base-pair duplex containing a centrally located 22-base-pair consensus binding sequence, the coumarin label being placed 22 residues from the 5′ end of the labeled strand. The solid line describes the best-fit relationship ($k_{AS} = 1.3 \times 10^8\ M^{-1}$) obtained by nonlinear regression analysis (M. F. Bailey and W. H. Sawyer, unpublished results). Dotted lines indicate the extrapolations used to determine the stoichiometry.

Even when there are no changes in the quantum yield or emission maximum of the fluorophore, fluorescence anisotropy can frequently be used to monitor the association reaction due to the slower rotational relaxation of the DNA–protein complex (Heyduk and Lee, 1990). An example of this approach is depicted in Figure 8.7, which summarizes results obtained for the binding of the TyrR repressor protein of *Escherichi coli* to a 42-base-pair duplex containing a centrally located 22-base-pair operator sequence.

B-DNA has a large and positive circular dichroism (CD) band in the region 260–280 nm. The aromatic amino acid residues of proteins do contribute to the CD in this region, but the contribution is weak (3%–5%) in comparison with that of DNA. Numerous studies have made use of this difference to monitor the specific and nonspecific binding of proteins to DNA (Fried *et al.*, 1983; Schnarr and Daune, 1984; Blazy *et al.*, 1987). Both increases and decreases in the magnitudes of this band as the result of protein binding have been noted, thereby indicating distortion of the DNA helix on association with protein. An example is shown in Figure 8.8 for the binding of single-stranded DNA binding protein from bacteriophage T4 (Kuil *et al.*, 1989), which clearly exhibits positive cooperativity (*cf.* Fig. 8.6) and has therefore been analyzed by the procedure (McGhee and von Hippel, 1974) outlined in Section 8.2.1.

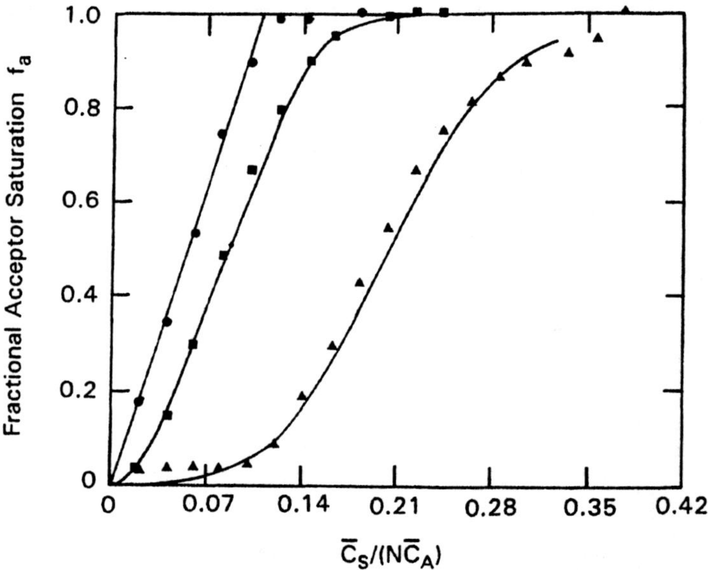

Fig. 8.8. The interaction of single-stranded DNA-binding protein GP 32 from bacteriophage T4 (S) with poly(rA), as monitored by the change in circular dichroism ellipticity at 260 nm that accompanies binding in experiments with 0.05 M (●), 0.35 M (■), and 0.45 M (▲) NaCl. The nucleotide concentration ($N\bar{C}_A$) of poly(rA) was 20–35 μM. [Adapted from Kuil *et al.* (1989).]

8.4. ANALYTICAL ULTRACENTRIFUGATION

The use of sedimentation equilibrium to characterize the binding of ligands (small or macromolecular) to an acceptor has been described in Section 4.3, where emphasis was placed on the use of the omega function (Nichol *et al.*, 1976) to evaluate the distribution of free ligand. Because a different approach has been adopted in the use of analytical ultracentrifugation to characterize DNA–protein interactions, a brief description of that procedure follows. The DNA used in sedimentation equilibrium studies is usually a short duplex DNA of up to 50–60 base pairs that contains a consensus sequence of 15–30 base pairs responsible for specific recognition by the DNA-binding protein. For example, repressor proteins frequently exist as dimers that bind to a palindromic operator sequence so that the binding stoichiometry is one repressor dimer per DNA duplex. The reaction may thus be considered in terms of the equilibrium $A + S \rightleftarrows AS$, where A is the DNA and S is the protein ligand.

At sedimentation equilibrium the two reactants and the complex thereof are present in chemical equilibrium at all points in the cell, their distributions being described by eq. 4.10. Provided that the total concentration of DNA (A and AS) can be monitored, the concentration distribution of the DNA constituent is given by the sum of the distributions of free DNA and DNA-protein complex. Thus, in terms of the free concentrations of A, $c_A(x_F)$, and AS, $c_{AS}(x_F)$ pertaining to a fixed radial distance, x_F, the total DNA concentration distribution, $\bar{c}_A(x)$, is given by

$$\bar{c}_A(x) = c_A(x_F) \exp [M_A(1 - \bar{v}_A\rho)\omega^2(x^2 - x_F^2)/2RT] + c_{AS}(x_F) \exp [M_{AS}(1 - \bar{v}_{AS}\rho)\omega^2(x^2 - x_F^2)/2RT] \quad (8.7)$$

The molecular weights of the DNA (M_A) and protein (M_S) are known or can be found from separate sedimentation equilibrium experiments, whereupon M_{AS} is obtainable as the sum of M_A and M_S. Similarly, $\bar{v}_{AS}$ may be obtained from the partial specific volumes of A and S. Specifically, $\bar{v}_{AS} = (M_A\bar{v}_A + M_S\bar{v}_S)/(M_A + M_S)$. The equilibrium distribution of the DNA constituent can therefore be fitted to solve for two unknowns, namely, $c_A(x_F)$ and $c_{AS}(x_F)$, the concentrations of the uncomplexed and complexed DNA species. Furthermore, because any radial distance within the sedimentation equilibrium distribution may be considered to be x_F, the net result of the curve-fitting procedure is delineation of the distribution of both DNA species throughout the equilibrium distribution.

The main problem associated with the use of absorption optics in the ultracentrifuge to study these interactions is the spectral overlap between the DNA and protein in the 260–280 nm region. Three strategies have been devised to overcome this problem. In the first, the interaction of the protein is studied with a relatively long stretch of DNA so that the nucleotide absorption at 260 nm blankets out any contribution from the protein. For example, when the *Escherichia coli* TyrR repressor protein interacts with a 42-base-pair duplex containing its 22-base-pair operator site, 96% of the absorption at 260 nm can be attributed to the DNA. The use of a long stretch of DNA also has advantages when flanking regions adjacent to the recognition site also contribute to the total binding energy of the interaction.

The second method is appropriate in situations where shorter DNA is used such that the protein contributes significantly to the absorption at 260 nm. The use of eq. 8.7 to fit data obtained at a single wavelength is clearly an ill-conditioned problem. However, the fitting of data collected at three or more wavelengths ranging from a region where the protein contribution dominates the absorption to a region where DNA is the dominant contributor to the absorption gives rise to an analysis for which the solution becomes highly constrained, whereupon the complete data set can be fitted by global procedures (Kim *et al.*, 1994).

The third approach involves monitoring the protein distribution in the presence of DNA. Because spectral overlap precludes the use of 280 nm for that purpose, genetic engineering needs to be employed to create a protein with its tryptophan residues replaced by 5-hydroxytryptophan, which absorbs at a wavelength (310 nm) where DNA absorption is negligible. This approach has been used successfully to study the association of the λ*cI* repressor with DNA (Laue *et al.*, 1993).

The use of the analytical ultracentrifuge will find major application in situations where the oligomeric state of the protein binding to the DNA is unknown. When the protein exists as an equilibrium mixture of monomer and higher polymer, preferential binding of one ligand state can lead to sigmoidal binding plots, the result being that the level of saturation of a specific DNA-binding site can be controlled over a relatively small range of protein concentration (Nichol and Winzor, 1972; Royer *et al.*, 1990). To that end, Laue *et al.*, (1993) have shown that under some conditions the λ*cI* repressor can bind to its operator as an octamer rather than a dimer and that the binding does not drive the octamer to a more dissociated state. However, the oligomeric equilibrium of the protein may be controlled by levels of a corepressor or effector molecule. For example, Fernando and Royer (1992) have reported that the presence of tryptophan favors the binding of dimeric rather than the tetrameric form of the *Escherichia coli trp* repressor to a 26-base-pair operator fragment. The oligomeric state of repressors may also permit the looping of DNA in situations where operators are sufficiently closely spaced on the DNA for binding at one site to effect cooperatively the expression at another (Ptashne, 1986; Gralla, 1989).

8.5. GEL RETARDATION ANALYSIS

As reported originally by Garner and Revzin (1981) and by Fried and Crothers (1981), the electrophoretic mobility of DNA in a nondenaturing polyacrylamide gel is decreased when complexed with protein. Two bands are revealed on staining for DNA—a faster band representing free DNA and a slower band representing the complex. A major consideration is whether the separation of free from bound species perturbs the thermodynamic equilibrium of the binding reaction. Two factors combine to decrease this possibility. First, for very strong interactions ($k_{AS} > 10^{12}\ \text{M}^{-1}$) the dissociation rate is sufficiently slow that little dissociation of the complex occurs within the duration of the electrophoretic experiment. Second, the cage effect of the surrounding gel polymer network decreases the rate of escape of dissociated reactants from the vicinity of the complex, thereby creating effectively elevated local concentrations of reactants and hence favoring complex formation. Nevertheless, some dissociation of complexes may occur during the course of the experiment, especially for weaker interactions, and hence lead to an underestimate of the concentration of complexed DNA. A phenomenological theory of gel electrophoresis (Cann, 1989, 1993) has contributed greatly to the utility of this procedure by providing a quantitative description of the cage effect and hence of the limiting conditions for validity of an analysis that relies upon the ability of the cage effect to prevent dissociation of DNA–protein complex(es) during the gel retardation assay.

Samples containing a constant concentration of DNA but varying concentrations of protein are subjected to gel electrophoresis in the usual way, and the distributions of DNA between free and protein-complexed states are quantified either by densitometry after staining or by excision and counting of radioactivity arising from ^{32}P-labeled DNA. The results are plotted as the dependence of the fraction of uncomplexed DNA upon protein concentration so that the binding constant may be determined from the reciprocal of the protein concentration at which half of the DNA is bound. A decided advantage of this method is that it can resolve complexes with different stoichiometries, because dual labeling techniques can be used to quantify the amounts of protein and DNA present in each complex. The error in the association constant for each liganded state depends upon the accuracy with which the concentrations of free and complexed species can be determined. In favorable cases it is possible to detect cooperativity of binding to multiple adjacent sites. Figure 8.9 provides an example of such an analysis applied to the binding of one,

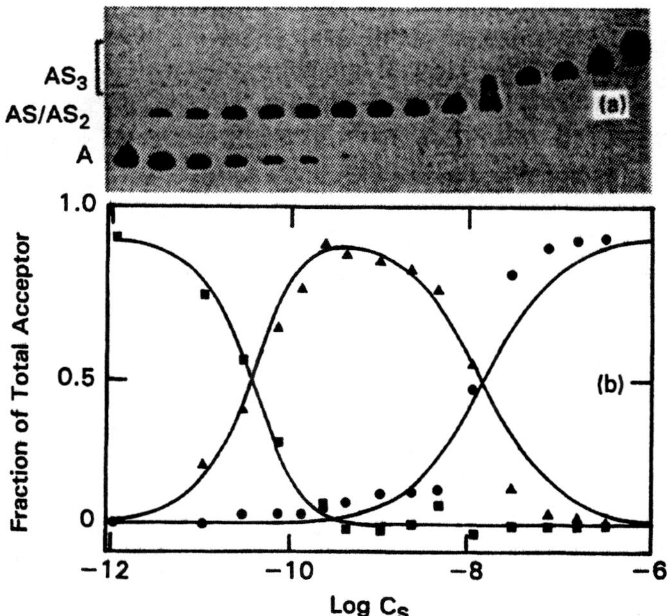

Fig. 8.9. Use of gel retardation assays to characterize the interaction of the λ*cI* repressor with a three-site operator. **(a)** The gel assay, in which the bands correspond to the unliganded operator (*A*) and complexes with one (*AS*), two (AS_2), and three (AS_3) or more repressor molecules bound. **(b)** Proportions of DNA molecules with zero (■), one or two (▲), and three (●) repressors bound. The theoretical curves are for a system with values of 2.24×10^{10} M^{-1}, 3.94×10^{20} M^{-2}, and 2.70×10^{28} M^{-3} for the equilibrium constants describing the formation of *AS*, AS_2, and AS_3 from the two reactants. [Adapted from Senear and Brenowitz (1991).]

two, or three λ*cI* repressors to operator DNA (Senear and Brenowitz, 1991). Practical and theoretical issues of gel retardation analysis have been reviewed by Carey (1991).

8.6. NITROCELLULOSE FILTER ASSAYS

Certain membrane filters are able to retain DNA–protein complexes but allow free DNA to pass through. The method was originally used to study the binding of RNA to ribosomes, but was modified subsequently for the study of DNA–protein equilibria (Jones and Berg, 1966). Usually, ^{32}P-labeled DNA is used so that the amount of complexed DNA, *i.e.*, the amount of DNA that is retained on the filter, can be determined directly from the radioactivity of the filter. Inasmuch as DNA concentrations of the order of 10^{-10} M can be used, the method is extremely sensitive. Reasonably accurate estimates of the binding constants for high-affinity interactions should, in principle, be obtained, because the rates of dissociation of such complexes are likely to be sufficiently slow to render negligible the amount of complex breakdown during the washing steps to remove the uncomplexed reactants. Nevertheless, retention efficiencies of 30%–40% at saturation are not unusual, possibly reflecting the loss of complex(es) during the washing steps. It is therefore important to determine whether the retention efficiency is constant throughout the binding curve (Riggs *et al.*, 1970b). Figure 8.10 shows the results of a filtration assay that were used to characterize the *lac* repressor-operator interaction, which has a k_{AS} of 6.7 × 10^{12} M^{-1} (Riggs *et al.*, 1970b). Other examples are provided by the λ repressor–operator interaction (Nelson and Sauer, 1985) and the *trp* repressor–operator interaction (Klig *et al.*, 1987).

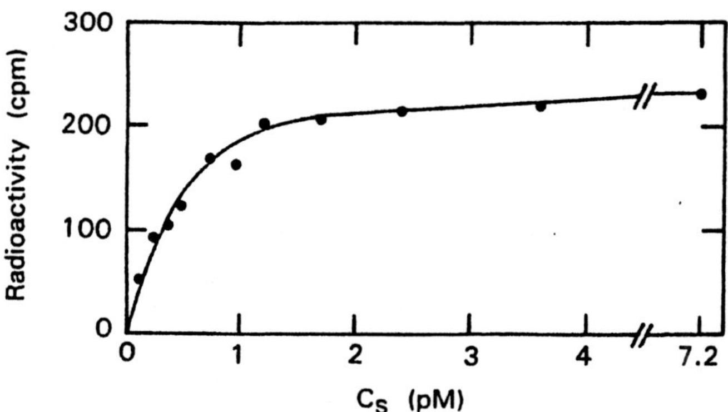

Fig. 8.10. Use of nitrocellulose filter assays to characterize the binding of the *lac* repressor to its operator in experiments in which the concentration of ^{32}P-labeled λφ80d*lac* DNA was fixed, that of the repressor being varied. The solid line is the theoretical dependence for an interaction with a binding constant of 6.7 × 10^{12} M^{-1}. [Adapted from Riggs *et al.* (1970b).]

8.7. INTERCALATION OF AROMATIC DYES

Some dyes intercalate between base pairs in DNA, thereby increasing the length of the polymer chain. In extreme cases the binding of a dye molecule at one m-residue site may totally exclude the binding of another molecule at adjacent sites, whereupon saturation of the polynucleotide entails the intercalation of a ligand molecule between every second base pair (Bauer and Vinograd, 1970; Bresloff and Crothers, 1975). This feature is illustrated in Figure 8.11 for the binding of ethidium bromide to circular SV40 DNA that had been nicked to render it a linear acceptor lattice. In keeping with the behavior associated with nonspecific binding, the Scatchard plot is of a negatively cooperative form; but the abscissa intercept of 0.25 signifies the involvement of four nucleotide residues in the binding of each ligand molecule despite the fact that ethidium bromide is intercalated between only two nucleotide residues. For an acceptor of infinite chainlength, the distinction between involvement and mere occupancy of residues in the binding process is immaterial, whereupon eq. 8.4 with $m = 4$ provides the necessary quantitative expression for evaluation of the intrinsic binding constant, k_{AS}. In that regard we note that Bauer and Vinograd (1970) derived a different quantitative expression, which was also based on a probability approach but entailed a more questionable approximation. Although studies of dye intercalation normally entail equilibrium binding measurements, the equilibrium constant for ethidium bromide binding has also been determined

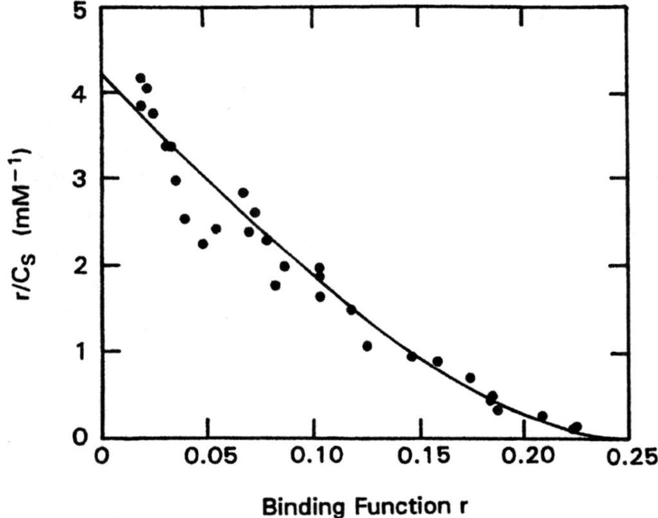

Fig. 8.11. Scatchard plot of binding data reflecting the intercalation of ethidium bromide into nicked circular SV40 DNA, the binding function being defined with the acceptor concentration expressed on a nucleotide basis. [Adapted from Bauer and Vinograd (1970).]

by measuring the association and dissociation rate constants by the pressure-jump technique (Macgregor *et al.*, 1985).

Bis and tris intercalators have also been synthesized in an effort to increase the affinity of the ligand for the DNA acceptor, there being a general tendency for the binding affinities and the number of bases associated with binding to increase with the number of intercalating groups within the one ligand molecule. For example, the number of base pairs entailed in the binding of acridine monomers is two, whereas covalently coupled acridine dimers and trimers occupy four and six base pairs, respectively, and the binding constant increases from 3×10^6 M^{-1} for the monomer to about 10^{14} M^{-1} for the trisacridine species (Laugaa *et al.*, 1985).

This excluded-site model also applies to those ligands that bind nonspecifically along the DNA helix. Thus the dye Hoescht 33258 binds snugly along the minor groove of the double helix; but in this instance the number of base pairs per binding site varies with base sequence (Loontiens *et al.*, 1990, 1991; Quintara *et al.*, 1991). Although this nonspecific binding to linear sequences of the DNA lattice again gives rise to negatively cooperative binding curves, the dependence of the size of the linear sequence upon base composition renders invalid its characterization via eq. 8.4. Those expressions reflect purely statistical considerations and hence the premise that binding is equally probable to all lattice sequences of a single prescribed length.

8.8. A WISHFUL LOOK INTO THE FUTURE

Although most current research in molecular biology is restricted to qualitative considerations of DNA–protein interactions, the above examples serve to illustrate the existence of a sizable selection of methods that may be applied to the quantification of the vast array of interactions involved in the regulation of gene expression. A major factor mitigating against such experimentation at the present moment seems to be a general lack of awareness within the cell biology research community of the need and importance to obtain equilibrium constants for DNA–protein interactions. The characterization of these interactions would, for example, allow investigation of the possibility that a combination of such equilibria and binding to preferred states may emerge as a common basic mechanism in the control of genetic expression and suppression of particular enzymatic/hormonal responses. At present the expression and suppression of such activities tend to be considered in terms of switch-on and switch-off mechanisms. It is therefore hoped that this chapter, taken in conjunction with its predecessors, may lead to a wider realization of the benefits to be derived from the quantitative characterization of DNA–ligand interactions, and also of some of the methodology already available for such quantification of nucleic acid interactions.

9

CONCLUDING REMARKS

The interaction between a ligand and an acceptor can be regarded as an equilibrium phenomenon that is described by the law of mass action. For reactions with 1 : 1 stoichiometry, that chemical equilibrium is manifested as a rectangular hyperbolic dependence of the concentration of complexed ligand upon its free concentration in experiments with a fixed total acceptor concentration. Such rectangular hyperbolic dependence of the binding function upon free ligand concentration is also observed in the binding of a univalent ligand to multiple acceptor sites that exhibit equivalence and independence of their interactions with ligand. Many binding studies are therefore relatively straightforward inasmuch as they merely entail the evaluation of the two binding parameters describing the rectangular hyperbolic dependence of r upon C_S, *viz.*, the intrinsic binding constant and the number of ligand-binding sites on the acceptor (or the effective total concentration of cellular receptor sites), via one or more of the procedures described in Chapters 2–5. Indeed, casual inspection of the literature may well convey the incorrect impression that the study of ligand binding should end at Chapter 5.

Despite the simplicity of the law of mass action, considerable caution must be exercised in its application to biological interactions because of complications that arise from unique features of either the ligand or the acceptor. For example, in many studies of binding to cellular receptors the potential for multivalence of a macromolecular ligand needs to be considered. Because an assumed univalence of protein ligands continues to pervade experimental receptor-binding studies, it is hoped that Section 7.1 may lead to more critical appraisal of the quantitative interpretation placed on experimental data for receptor–protein interactions. The question of ligand multivalence has not yet been addressed in studies of DNA–protein interactions, which are often already complicated by the binding of ligand to random nucleotide sequences as well as to the specific operator sequence of the DNA. In that field adequate attention has certainly been given to the effects of such nonspecific binding on the thermodynamic characterization of DNA–ligand interactions (Chapter 8). However, as noted in Section 8.1, considerable patience may be required in order to obtain experimental results that reflect true equilibrium, and that are therefore amenable to interpretation in terms of those thermodynamic expressions. The other major complication that is frequently encountered in binding studies is an interplay of interactions whereby the concentration of a given form of acceptor, ligand, or acceptor–ligand complex reflects involvement of the species in two competing equilibria (Chapter 6). In the sense that such interplays of equilibria can give rise to the sigmoidal responses that are central to regulation at the enzymatic, metabolic, hormonal, and genetic levels, it becomes imperative that a quantitative description of the phenomenon be available if we are to understand the subtleties of biological control in molecular terms.

From the viewpoint of characterizing acceptor–ligand interactions *in vitro*, X-ray crystallography and NMR techniques are providing an ever-increasing number of high-resolution acceptor–ligand structures. Our capacity to understand the thermodynamics of binding in terms of the specific inter- and intramolecular bonds has thus improved greatly and will presumably continue to do so at an increasing rate. However, static molecular structures by themselves do not comment upon the dynamic nature of complex formation, which is the key aspect of acceptor–ligand interactions in regard to biological control. Thus, although such structures may provide invaluable information on the numbers of binding sites for a ligand or an effector, they do not address the dependence of complex formation upon the concentrations of reactants—factors that are of primary importance to our understanding of the extent and consequences of the interaction in a system under defined conditions (specified reactant concentrations). Chapter 6 has described ways of identifying and quantifying many of the mechanisms that can control ligand binding—cooperativity of acceptor sites, isomerization or self-association of either the acceptor or the ligand, and the binding of effector molecules.

Although control mechanisms involving an interplay of equilibria have been shown to exist in purified systems *in vitro*, their relevance *in vivo* remains to be assessed in most instances. However, the conservation of such mechanisms throughout evolution is certainly consistent with their involvement in the control of enzyme function, of transport kinetics, and of metabolism in general. Establishment of the importance of these interplays of equilibria as regulating mechanisms *in vivo* thus remains an important challenge, this being a goal that must, however, await the development of procedures for taking into account the consequences of thermodynamic nonideality in the crowded molecular environment of the cell cytoplasm (Section 7.5).

The sentiments expressed in the preceding paragraphs are not intended to be either an admission of defeat or a deterrant to those contemplating an adventure into the study of acceptor–ligand interactions. Instead, they are intended to convey the impression that the quantitative characterization of acceptor–ligand interactions will have to become an important aspect of biochemistry, molecular biology, pharmacology, immunology, and neurobiology if we are ever to gain a complete understanding of physiological systems at the molecular level. The task is enormous, and its completion date undoubtedly a long way into the future. Nevertheless, significant advances have been made in the five decades since Klotz (1946) provided the quantitative explanation of a rectangular hyperbolic dependence of binding responses upon free ligand concentration for multivalent acceptors. That landmark publication has triggered interest not only in the development of methods for the quantitative characterization of ligand binding, but also in the formulation of potential mechanisms of biological regulation. In this book we have attempted to combine these two interests by outlining ways to maximize the amount of meaningful mechanistic information that can be gleaned from well-conceived quantitative studies of ligand binding. Because it attracted the attention of physical chemists, the development of methodology for the evaluation of binding parameters is at a relatively advanced stage compared with its application to the quantitative characterization of the myriads of acceptor–ligand interactions encountered in biology. The field is thus ripe for takeover by the biologists, whose knowledge of the intricacies of a particular biological acceptor–ligand interaction is clearly going to facilitate greatly its quantitative characterization under the most appropriate conditions. In the past this knowledge has often been used to advantage in the selection of experimental conditions for the study of acceptor–ligand interactions; but, unfortunately, the significance of much of the quantitative information derived has been questionable because inadequate attention has been paid to the design of the binding experiments and/or the interpretation of the experimental results. May this book help to correct that tendency.

This book has emphasized that the quantitative characterization of an acceptor–ligand system can demand considerable intellectual input—a requirement that has always been taken for granted in basic scientific endeavors but that is seldom necessary for most projects in molecular biotechnology. It is hoped that the publication of this book may stimulate interest in the quantitative characterization of acceptor–ligand interactions and thus be a catalyst for major breakthroughs in our understanding of the molecular events responsible for the regulation of cell physiology.

GLOSSARY OF SYMBOLS

A	acceptor molecule
a_i	thermodynamic activity of species i
$B_{i,j}$	second virial coefficient for the mutual interactions of species i and j
c_i	free concentration (g/unit volume) of species i
$\bar{c}_i$	total concentration (g/unit volume) of species i
C_i	free concentration (molar) of species i
$\bar{C}_i$	total concentration (molar) of species i in liquid phase
$\bar{\bar{C}}_i$	total concentration (molar) of species i in a two-phase system
D	dielectric constant
e	electronic charge
E	effector molecule
f	number of sites for acceptor on ligand (ligand valence)
F	inert space-filling molecule
f_a	fractional saturation of acceptor sites by ligand
f_s	fractional saturation of ligand by acceptor
$\mathbf{G}$	Gibbs free energy
$\mathbf{H}$	enthalpy
I	inhibitor in competitive binding assays
$\mathbf{I}$	ionic strength
k_d	intrinsic dissociation constant (reciprocal of k_{ij})
k_f, k_r	association and dissociation rate constants for acceptor–ligand complex formation

k_{ij} intrinsic binding constant (M^{-1}) for interaction between sites on species *i* and *j*

$\bar{k}_{AX}$ constitutive equilibrium constant (M^{-1}) for the solute–matrix interaction in competitive, solid-phase assays

K_1, K_2 stepwise binding constants (M^{-1}) for successive attachments of ligand to acceptor

K_m Michaelis constant for an enzyme-catalyzed reaction

K_p partition coefficient describing solute partition between organic and aqueous phases

$\mathbf{K}_{ij}$ equilibrium constant for the formation of the complex AS_iE_j from the individual reactants *A*, *S*, and *E*

L, *L'* forms of ligand coexisting in isomerization or self-association equilibrium

m length of residue sequence of a polymer occupied by a ligand molecule

M monomeric form of an acceptor undergoing self-association

M_i molecular weight of species *i*

$\overline{M}_w$ weight-average molecular weight

n stoichiometry of acceptor or ligand self-association

N number of residues comprising a polymer chain

N Avogadro' number

p number of sites for ligand on an acceptor (*A*), the monomeric form of an acceptor (*M*), or the *R* form of an isomerizing acceptor

P polymeric form of an acceptor undergoing self-association

q number of ligand-binding sites on the polymeric form of an acceptor, or the *T* state of an isomerizing acceptor

Q ratio of solute–matrix binding constants ($k_{AX}/\bar{k}_{AX}$) measured in the absence and presence of a competing ligand

r binding function for a univalent ligand (eq. 1.3)

R isomeric form of acceptor in equilibrium coexistence with form *T*

R universal gas constant

r_f binding function of an *f*-valent ligand (eq. 7.1)

R_i effective spherical radius of hydrated species *i*

R_p plateau response in studies of ligand binding by biosensor technology

S ligand molecule

S entropy

T isomeric form of acceptor in equilibrium coexistence with form *R*

T temperature (absolute)

$U_{i,j}$ covolume of species *i* and *j* (eq. 7.23)

v initial velocity of enzyme-catalyzed reaction

v_i velocity of migration of pure species *i*

$\overline{v}_i$	partial specific volume of species i, also constituent velocity of species i (Chapter 4)
v_m	maximal initial velocity of an enzyme-catalyzed reaction
V_i	elution volume of species i
$\overline{V}_i$	constituent elution volume of species i
V_i^*	volume accessible to species i
w	number of sites on R isomer for effector E (Chapter 6); also the electrostatic interaction factor (Chapter 4)
x	radial distance in sedimentation equilibrium
x_F	reference radial distance in sedimentation equilibrium
X	matrix-bound reactant in solid-phase assays (including affinity chromatography); also the equilibrium constant (M^{1-n}) describing pre-existing acceptor isomerization or self-association
X^*	constitutive self-association constant
y	measured spectroscopic parameter
y_b	spectroscopic parameter for solute (acceptor or ligand) in bound state
y_f	spectroscopic parameter for solute (acceptor or ligand) in free state
Y	equilibrium constant for ligand-induced isomerization or self-association of acceptor
z	number of sites on T isomer for acceptor
Z_i	net electrostatic charge (valence) or species i
α	free fraction ($C_{S*}/\overline{C}_{S*}$) of radiolabeled antigen in radioimmunoassays
β	free fraction of a competitive inhibitor of radiolabeled antigen in radioimmunoassays
γ_i	activity coefficient of species i
κ	inverse screening length
Ξ	grand partition function (Wyman linked-functions theory)
ω	cooperativity factor in ligand binding, also angular velocity in centrifugation
$\Omega_i(x)$	omega function for species i at radial distance x (eq. 4.12)

REFERENCES

Ackers, G. K., and Thompson, T. E. (1965) *Proc. Natl. Acad. Sci. USA* **53,** 342.

Adair, G. S (1925) *J. Biol. Chem.* **63,** 529.

Albertsson, P.-A., and Philipson, L. (1960) *Nature* **185,** 38.

Allen, D. W., Guthe, K. F., and Wyman, J. (1950) *J. Biol. Chem.* **187,** 393.

Alma, N. C. M., Harmsen, B. J. M., de Jong, E. A. M., Ven, J. V. D., and Hilbers, C. W. (1983) *J. Mol. Biol.* **163,** 47.

Anderson, S. R., and Weber, G. (1965) *Biochemistry* **4,** 1948.

Andrews, P., Kitchen, B. J., and Winzor, D. J. (1973) *Biochem. J.* **135,** 897.

Aune, K. C., and Timasheff, S. N. (1971) *Biochemistry* **10,** 1609.

Baghurst, P. A., and Nichol, L. W. (1975) *Biochim. Biophys. Acta* **412,** 168.

Baghurst, P. A., Nichol, L. W., and Winzor, D. J. (1978) *J. Theor. Biol.* **74,** 523.

Bailey, I. A., Williams, S. R., Radda, G. K., and Gadian, D. G. (1981) *Biochem. J.* **196,** 171.

Bates, R. G. (1964) *Determination of pH: Theory and Practice*, 2nd ed. Wiley, New York.

Bauer, W., and Vinograd, J. (1970) *J. Mol. Biol.* **47,** 419.

Bennett, W. S., and Steitz, T. A. (1980) *J. Mol. Biol.* **140,** 211.

Berde, C. B., Nagai, M., and Deutsch, H. F. (1979) *J. Biol. Chem.* **254,** 12609.

Berg, D. G., Winter, R. B., and von Hippel, P. H. (1982) *Trends Biochem. Sci.* **7,** 52.

Bergman, D. A., and Winzor, D. J. (1986) *Anal. Biochem.* **153,** 380.

Bergman, D. A., and Winzor, D. J. (1989a) *Eur. J. Biochem.* **185,** 91.

Bergman, D. A., and Winzor, D. J. (1989b) *J. Theor. Biol.* **137,** 171.

Bergman, D. A., Shearwin, K. E., and Winzor, D. J. (1989) *Arch. Biochem. Biophys.* **274,** 55.

Berliner, L. J. (1981) in *Spectroscopy in Biochemistry* (Bell, J. E., ed.), vol. 2, p. 1. CRC Press, Boca Raton, FL.

Berliner, L. J., and Wong, S. S. (1975) *Biochemistry* **14,** 4977.

Blatt, W. F., Robinson, S. M., and Bixler, H. J. (1968) *Anal. Biochem.* **26,** 151.

Blazy, B., Culard, F., and Maurizot, J. C. (1987) *J. Mol. Biol.* **195,** 175.

Bogoyevitch, M. A., Gillam, E. M. J., Reilly, P. E. B., and Winzor, D. J. (1987) *Biochem. Pharmacol.* **36,** 4167.

Booth, C. K., Nixon, P. F., and Winzor, D. J. (1992) *J. Chromatogr.* **609,** 83.

Boschelli, F., Arndt, K., Nick, H., Zhang, Q., and Li, P. (1982) *J. Mol. Biol.* **162,** 251.

Bothwell, M. A., Howlett, G. J., and Schachman, H. K. (1978) *J. Biol. Chem.* **253,** 2073.

Brandts, J. F., and Lin, L.-N, (1990) *Biochemistry* **29,** 6927.

Bresloff, J. L., and Crothers, D. M. (1975) *J. Mol. Biol.* **95,** 103.

Briggs, G. E., and Haldane, J. S. (1925) *Biochem. J.* **19,** 338.

Brinkworth, R. I., Masters, C. J., and Winzor, D. J. (1975) *Biochem. J.* **151,** 631.

Brocklebank, A. M., Sawyer, W. H., Wiley, J. S., and Winzor, D. J. (1993) *Anal. Biochem.* **213,** 104.

Bujalowski, W., and Lohman, T. M. (1987) *J. Mol. Biol.* **195,** 897.

Calvert, P. D., Nichol, L. W., and Sawyer, W. H. (1979) *J. Theor. Biol.* **80,** 233.

Cann, J. R. (1989) *J. Biol. Chem.* **264,** 17032.

Cann, J. R. (1993) *Electrophoresis* **14,** 669.

Cann, J. R., and Winzor, D. J. (1987) *Arch Biochem. Biophys.* **256,** 78.

Cann, J. R., Nichol, L. W., and Winzor, D. J. (1981) *Mol. Pharmacol.* **20,** 244.

Cann, J. R., Appu Rao, A. G., and Winzor, D. J. (1989) *Arch. Biochem. Biophys.* **270,** 173.

Cann, J. R., Coombs, R. O., Howlett, G. J., Jacobsen, M. P., and Winzor, D. J. (1994) *Biochemistry* **33,** 10185.

Cantor, C. R., and Schimmel, P. R. (1980) *Biophysical Chemistry.* Freeman, San Francisco.

Carey, J. (1991) *Methods Enzymol.* **208,** 103.

Cecil, R., and Robinson, G. B. (1975) *Biochim. Biophys. Acta* **404,** 164.

Changeux, J.-P., and Rubin, M. M. (1968) *Biochemistry* **7,** 553.

Chanutin, A., Ludewig, S., and Masket, A. V. (1942) *J. Biol. Chem.* **143,** 737.

Chatelier, R. C., and Sawyer, W. H. (1987) *J. Biochem. Biophys. Methods* **15,** 49.

Chatelier, R. C., Ashcroft, R. G., Lloyd, C. J., Nice, E. C., Whitehead, R. H., Sawyer, W. H., and Burgess, A. W. (1986) *EMBO J.* **5,** 1181.

Colman, R. F. (1972) *Anal. Biochem.* **46,** 358.

Colowick, S. P., and Womack, F. C. (1969) *J. Biol. Chem.* **244,** 774.

Conway, A., and Koshland, D. E., Jr. (1968) *Biochemistry* **7,** 4011.

Cooper, P. F., and Wood, G. C. (1968) *J. Pharm. Pharmacol.* **20,** 150S.

Cumps, J., Razzouk, C., and Roberfroid, M. B. (1977) *Chem.-Biol. Interact.* **16,** 23.

Cush, R., Cronin, J. M., Stewart, W. J., Maule, C. H., Molloy, J., and Goddard, N. J. (1993) *Biosensors Bioelectronics* **8,** 347.

Dalziel, K. (1968) *FEBS Lett.* **1,** 346.

De Cristofaro, R., Landolfi, R., and Di Cera, E. (1990) *Biophys. Chem.* **36,** 77.

De Leo, D. T., and Helmerhorst, E. (1992) *Anal. Biochem.* **206,** 207.

Denburg, J., and De Luca, M. (1968) *Biochem. Biophys. Res. Commun.* **31,** 453.

Dombrose, F. A., Gitel, S. N., Zawalich, K., and Jackson, C. M. (1979) *J. Biol. Chem.* **254,** 5027.

Drewe, R. H., and Winzor, D. J. (1976) *Biochem. J.* **159,** 737.

Dunn, B. M., and Chaiken, I. M. (1974) *Proc. Natl. Acad. Sci. USA* **71,** 2382.

Dunn, B. M., and Chaiken, I. M. (1975) *Biochemistry* **14,** 2343.

Dunn, B. M., Danner-Rabovsky, J., and Cambias, J. S. (1983) in *Affinity Chromatography and Biological Recognition* (Chaiken, I. M., Wilchek, M., and Parikh, I., eds.), p. 93. Academic Press, New York.

Dwek, R. A. (1973) *Nuclear Magnetic Resonance in Biochemistry.* Clarendon Press, Oxford.

Edmond, E., and Ogston, A. G. (1968) *Biochem. J.* **109,** 569.

Edsall, J. T., and Gutfreund, H. (1983) *Biothermodynamics: The Study of Biochemical Processes at Equilibrium.* Wiley, Chichester.

Edwards, K., Chan, R. Y. S., and Sawyer, W. H. (1994) *Biochemistry* **33,** 13304.

Elling, L., Kula, M.-R., Hadas, E., and Katchalski-Katzir, E. (1991) *Anal. Biochem.* **192,** 74.

Epstein, I. R. (1978) *Biophys. Chem.* **8,** 327.

Epstein, I. R. (1979) *Biopolymers* **18,** 2037.

Estabrook, R. W., Peterson, J., Baron, J., and Hildebrandt, A. (1972) in *Methods in Pharmacology* (Chignell, C. F., ed.), vol. 2, p. 303. Appleton-Century-Crofts, New York.

Fairclough, G. F., Jr., and Fruton, J. S. (1966) *Biochemistry* **5,** 673.

Ferdinand, W. (1964) *Biochem. J.* **92,** 578.

Ferguson, W. E., Smith, C. W., Adams, E. T., Jr., and Barlow, G. H. (1974) *Biophys. Chem.* **1,** 325.

Fernando, T., and Royer, C. A. (1992) *Biochemistry* **31,** 3429.

Ford, C. L., and Winzor, D. J. (1981) *Anal. Biochem.* **114,** 146.

Ford, C. L., Winzor, D. J., Nichol, L. W., and Sculley, M. J. (1984) *Biophys. Chem.* **18,** 1.

Freire, E., Mayorga, O. L., and Straume, M. (1990) *Anal. Chem.* **62,** 950A.

Fried, M. G., and Crothers, D. M. (1981) *Nucleic Acids Res.* **9,** 6505.

Fried, M. G., Wu, H.-M., and Crothers, D. M. (1983) *Nucleic Acids Res.* **11,** 2479.

Frieden, C. (1964) *J. Biol. Chem.* **239,** 3522.

Frieden, C. (1967) *J. Biol. Chem.* **242,** 4045.

Frieden, C., and Colman, R. F. (1967) *J. Biol. Chem.* **242,** 1705.

Gardner, J. A., and Mathews, K. S. (1990) *J. Biol. Chem.* **265,** 21061.

Garner, M. M., and Revzin, A. (1981) *Nucleic Acids Res.* **9,** 3047.

Gerhart, J. C. (1970) *Curr. Topics Cell. Regul.* **2,** 275.

Gilbert, G. A. (1966) *Nature* **210,** 299.

Gilbert, G. A., and Jenkins, R. L. (1959) *Proc. R. Soc. (Lond.)* **A253,** 420.

Gilbert, G. A., and Kellett, G. L. (1971) *J. Biol. Chem.* **246,** 6079.

Goldberg, R. J. (1952) *J. Am. Chem. Soc.* **74,** 5715.

Goldberg, R. J. (1953) *J. Am. Chem. Soc.* **75,** 3127.

Goodman, D. S. (1958a) *J. Am. Chem. Soc.* **80,** 3887.

Goodman, D. S. (1958b) *J. Am. Chem. Soc.* **80,** 3892.

Gow, A., Winzor, D. J., and Smith, R. (1990) *J. Theor. Biol.* **145,** 407.

Gralla, J. D. (1989) *Cell* **57,** 193.

Griffin, J. H., Schechter, A. N., and Cohen, J. S. (1973) *Ann. N.Y. Acad. Sci.* **222,** 693.

Guest, C. R., Hochstrasser, R. A., Dupuy, C. G., Allen, D. J., Benkovic, S. J., and Millar, D. P. (1991) *Biochemistry* **30,** 8759.

Haigh, E. A., and Sawyer, W. H. (1978) *Aust. J. Biol. Sci.* **31,** 1.

Hammond, G. L., Nisker, J. A., Jones, L. A., and Siiteri, P. K. (1980) *J. Biol. Chem.* **255,** 5023.

Hard, T., Sayre, M. H., Geiduschek, E. P., and Kearns, D. R. (1989) *Biochemistry* **28,** 2813.

Harris, S. J., and Winzor, D. J. (1985) *Arch. Biochem. Biophys.* **243,** 598.

Harris, S. J., and Winzor, D. J. (1987) *Biochim. Biophys. Acta* **911,** 121.

Harris, S. J., and Winzor, D. J. (1988a) *Anal. Biochem.* **169,** 319.

Harris, S. J., and Winzor, D. J. (1988b) *Arch. Biochem. Biophys.* **265,** 458.

Harris, S. J., Jackson, C. M., and Winzor, D. J. (1995a) *Arch. Biochem. Biophys.* **316,** 20.

Harris, S. J., Jackson, C. M., and Winzor, D. J. (1995b) *J. Protein Chem.* (in press).

He, J. J., and Mathews, K. S. (1990) *J. Biol. Chem.* **265,** 731.

Hearon, J. Z., Bernhard, S. A., Freis, S. L., Botts, J. D., and Morales, M. F. (1959) in *The Enzymes* (Boyer, P. D., Lardy, H., and Myrbäck, K., eds.), vol. 1, p. 49. Academic Press, New York.

Heidelberger, M. (1939) *Bacteriol. Rev.* **3,** 49.

Heidelberger, M., and Kendall, F. E. (1935) *J. Exp. Med.* **61,** 563.

Heyduk, T., and Lee, J. C. (1980) *Proc. Natl. Acad. Sci. USA* **87,** 1744.

Heyn, M. P., and Bretz, R. (1965) *J. Phys. Chem.* **69,** 3872.

Hinman, N. D., and Cann, J. R. (1976) *Mol. Pharmacol.* **12,** 769.

Hogg, P. J., and Winzor, D. J. (1984) *Arch. Biochem. Biophys.* **234,** 55.

Hogg, P. J., and Winzor, D. J. (1985) *Biochim. Biophys. Acta* **843,** 159.

Hogg, P. J., and Winzor, D. J. (1987a) *Anal. Biochem.* **163,** 331.

Hogg, P. J., and Winzor, D. J. (1987b) *Arch. Biochem. Biophys.* **254,** 92.

Hogg, P. J., Johnston, S. C., Bowles, M. R., Pond, S. M., and Winzor, D. J. (1987a) *Mol. Immunol.* **24,** 797.

Hogg, P. J., Reilly, P. E. B., and Winzor, D. J. (1987b) *Biochemistry* **26,** 1867.

Hogg, P. J., Jackson, C. M., and Winzor, D. J. (1991) *Anal. Biochem.* **192,** 303.

Holbrook, J. J. (1972) *Biochem. J.* **128,** 921.

Holbrook, J. J., Yates, D. W., Reynolds, S. J., Evans, R. W., Greenwood, C., and Gore, M. G. (1972) *Biochem. J.* **128,** 933.

Horbett, T. A., and Teller, D. C. (1973) *Biochemistry* **12,** 1349.

Hummel, J. P., and Dreyer, W. J. (1962) *Biochim. Biophys. Acta* **46,** 358.

Imai, K., Morimoto, H., Kotani, M., Watari, H., Hirata, W., and Kuroda, M. (1970) *Biochim. Biophys. Acta* **200,** 189.

Jacobsen, M. P., and Winzor, D. J. (1995) *Biochim. Biophys. Acta* **1246,** 17.

James, T. L., and Noggle, J. H. (1969) *Proc. Natl. Acad. Sci. USA* **62,** 644.

Jameson, D. M., and Sawyer, W. H. (1995) *Methods Enzymol.* **246,** 283.

Jarvis, T. C., Ring, D. M., Daube, S. S., and von Hippel, P. H. (1990) *J. Biol. Chem.* **265,** 15160.

Jeffrey, P. D., Nichol, L. W., and Teasdale, R. D. (1979) *Biophys. Chem.* **10,** 379.

Johnston, S. C., Bowles, M., Winzor, D. J., and Pond, S. M. (1988) *Fundam. Appl. Toxicol.* **11,** 261.

Jones, O. W., and Berg, P. (1966) *J. Mol. Biol.* **22,** 199.

Kalinin, N. L., Ward, L. D., and Winzor, D. J. (1995) *Anal. Biochem.* (in press).

Karlsson, R., Michaelson, R., and Mattson, L. (1991) *J. Immunol. Methods* **145,** 229.

Kim, S.-J., Tsukiyama, T., Lewis, M. S., and Wu, C. (1994) *Protein Sci.* **3,** 1040.

Klig, L. S., Crawford, I. P., and Yanofsky, C. (1987) *Nucleic Acids Res.* **15,** 5339.

Klotz, I. M. (1946) *Arch. Biochem.* **9,** 109.

Klotz, I. M. (1982) *Science* **217,** 1247.

Klotz, I. M. (1983) *Science* **220,** 981.

Koshland, D. E., Jr., and Neet, K. E. (1968) *Annu. Rev. Biochem.* **37,** 359.

Koshland, D. E., Jr., Némethy, G., and Filmer, D. (1966) *Biochemistry* **5,** 365.

Kowalczykowski, S. C., Lonberg, N., Newport, J. W., and von Hippel, P. H. (1981) *J. Mol. Biol.* **145,** 75.

Kowalczykowski, S. C., Paul, L. S., Lonberg, N., Newport, J. W., McSwiggen, J. A., and von Hippel, P. H. (1986) *Biochemistry* **25,** 1226.

Kuchel, P. W., Nichol, L. W., and Jeffrey, P. D. (1974) *J. Theor. Biol.* **48,** 39.

Kuil, M. E., van Amerongen, H., van der Vliet, P. C., and van Grondelle, R. (1989) *Biochemistry* **28,** 9795.

Kuter, M. R., Masters, C. J., and Winzor, D. J. (1983) *Arch. Biochem. Biophys.* **225,** 384.

Kyprianou, P., and Yon, R. J. (1982) *Biochem. J.* **207,** 549.

Lai, C. Y., and Horecker, B. L. (1972) *Essays Biochem.* **8,** 149.

Lakowicz, J. R. (1983) *Principles of Fluorescence Spectroscopy.* Plenum Press, New York.

Lamm, M. E., and Neville, D. M. (1965) *J. Phys. Chem.* **69,** 3872.

Lanir, A., and Navon, G. (1971) *Biochemistry* **10,** 1024.

Latt, A. L., and Sober, H. A. (1967) *Biochemistry* **6,** 3293.

Laue, T. M., Senear, D. F., Eaton, S., and Ross, J. B. A. (1993) *Biochemistry* **32,** 2469.

Laugaa, P., Markovits, J., Delbarre, A., Le Pecq, J.-B., and Roques, B. P. (1985) *Biochemistry* **24,** 5567.

Lin, L.-N., Mason, A. B., Woodworth, R. C., and Brandts, J. F. (1991) *Biochemistry* **30,** 11660.

Linderstom-Lang, K. (1924) *Compt. rend. trav. lab. Carlsberg* **15,** No. 7.

Löfås, S., and Johnsson, B. (1990) *J. Chem. Soc. Chem. Commun.* 1526.

Lohman, T. M., and Mascotti, D. P. (1992) *Methods Enzymol.* **212,** 424.

Lohman, T. M., Overman, L. B., and Datta, S. (1986) *J. Mol. Biol.* **187,** 603.

Loontiens, F. G., Regenfuss, P., Zechel, A., Dumortier, L., and Clegg, R. M. (1990) *Biochemistry* **29,** 9029.

Loontiens, F. G., McLaighlin, L. W., Diekman, S., and Clegg, R. M. (1991) *Biochemistry* **30,** 182.

Luttrell, B. M. (1993) *J. Biol. Chem.* **268,** 1521.

Macgregor, R. B., Clegg, R. M., and Jovin, T. M. (1985) *Biochemistry* **24,** 5503.

Masters, C. J., Sheedy, R. J., Winzor, D. J., and Nichol, L. W. (1969) *Biochem. J.* **112,** 806.

McGhee, J. D., and von Hippel, P. H. (1974) *J. Mol. Biol.* **86,** 449.

McMillan, W. G., and Mayer, J. E. (1945) *J. Chem. Phys.* **13,** 276.

Meadows, D. H., and Jardetzky, O. (1968) *Proc. Natl. Acad. Sci. USA* **61,** 406.

Meadows, D. H., Jardetsky, O., Epand, R. M., Ruterjans, H. H., and Scheraga, H. A. (1968) *Proc. Natl. Acad. Sci. USA* **60,** 766.

Michaelis, L., and Menten, L. M. (1913) *Biochem. Z.* **49,** 333.

Mills, F. C., Johnson, M. L., and Ackers, G. K. (1976) *Biochemistry* **15,** 5350.

Milthorpe, B. K., Jeffrey, P. D., and Nichol, L. W. (1975) *Biophys. Chem.* **3,** 169.

Minton, A. P. (1983) *Mol. Cell. Biochem.* **55,** 119.

Minton, A. P., and Wilf, J. (1981) *Biochemistry* **20,** 4821.

Monod, J., Wyman, J., and Changeux, J.-P. (1965) *J. Mol. Biol.* **12,** 88.

Müller, R. (1980) *J. Immunol. Methods* **34,** 345.

Müller, R. (1983) *Methods Enzymol.* **92,** 589.

Munro, P. J., Winzor, D. J., and Cann, J. R. (1993) *J. Chromatogr.* **659,** 267.

Murthy, M. R. N., Garavito, R. M., Johnson, J. E., and Rossman, M. G. (1980) *J. Mol. Biol.* **138,** 859.

Nelson, H. C. M., and Sauer, R. T. (1985) *Cell* **42,** 549.

Newport, J. W., Lonberg, N. K., Kowalczykowski, S. C., and von Hippel, P. H. (1981) *J. Mol. Biol.* **145,** 105.

Nichol, L. W., and Winzor, D. J. (1964) *J. Phys. Chem.* **68,** 2455.

Nichol, L. W., and Winzor, D. J. (1972) *Migration of Interacting Systems*, p. 18. Clarendon Press, Oxford.

Nichol, L. W., and Winzor, D. J. (1976) *Biochemistry* **15,** 3015.

Nichol, L. W., and Winzor, D. J. (1981) in *Protein–Protein Interactions* (Frieden, C., and Nichol, L. W., eds.), p. 337. Wiley, New York.

Nichol, L. W., Jackson, W. J. H., and Winzor, D. J. (1967a) *Biochemistry* **6,** 2449.

Nichol, L. W., Ogston, A. G., and Winzor, D. J. (1967b) *Arch. Biochem. Biophys.* **121,** 517.

Nichol, L. W., Ogston, A. G., and Winzor, D. J. (1967c) *J. Phys. Chem.* **71,** 726.

Nichol, L. W., Smith, G. D., and Ogston, A. G. (1969) *Biochim. Biophys. Acta* **184,** 1.

Nichol, L. W., Jackson, W. J. H., and Smith, G. D. (1971a) *Arch. Biochem. Biophys.* **144,** 438.

Nichol, L. W., Sawyer, W. H., and Winzor, D. J. (1971b) *Biochem. J.* **112,** 259.

Nichol, L. W., Jackson, W. J. H., and Winzor, D. J. (1972a) *Biochemistry* **11,** 585.

Nichol, L. W., O'Dea, K., and Baghurst, P. A. (1972b) *J. Theor. Biol.* **34,** 255.

Nichol, L. W., Kuchel, P. W., and Jeffrey, P. D. (1974a) *Biophys. Chem.* **2,** 354.

Nichol, L. W., Ogston, A. G., Winzor, D. J., and Sawyer, W. H. (1974b) *Biochem. J.* **143,** 435.

Nichol, L. W., Jeffrey, P. D., and Milthorpe, B. K. (1976) *Biophys. Chem.* **4,** 259.

Nichol, L. W., Wills, P. R., and Winzor, D. J. (1979) *J. Theor. Biol.* **80,** 39.

Nichol, L. W., Ogston, A. G., and Wills, P. R. (1981a) *FEBS Lett.* **126,** 18.

Nichol, L. W., Ward, L. D., and Winzor, D. J. (1981b) *Biochemistry* **20,** 4856.

Nichol, L. W., Sculley, M. J., and Winzor, D. J. (1982) *J. Theor. Biol.* **96,** 723.

Nichol, L. W., Sculley, M. J., Ward, L. D., and Winzor, D. J. (1983) *Arch. Biochem. Biophys.* **222,** 574.

Nichol, L. W., Owen, E. A., and Winzor, D. J. (1985) *Arch. Biochem. Biophys.* **239,** 147.

Nicolas, P., Camier, M., Dessen, P., and Cohen, P. (1976) *J. Biol. Chem.* **251,** 3965.

Nicolas, P., Dessen, P., Camier, M., and Cohen, P. (1978) *FEBS Lett.* **86,** 188.

Nowak, T. (1981) in *Spectroscopy in Biochemistry* (Bell, J. E., ed.), vol. 2, p. 109. CRC Press, Boca Raton, FL.

Nozaki, Y., and Tanford, C. (1967) *J. Am. Chem. Soc.* **89,** 742.

Olson, S. T., Halvorson, H. R., and Björk, I. (1991) *J. Biol. Chem.* **266,** 6342.

O'Shannessy, D. J., Brigham-Burke, M., Soneson, K. K., Hensley, P., and Brooks, I. (1993) *Anal. Biochem.* **212,** 457.

Parker, F. S., and Bhaskar, K. R. (1968) *Biochemistry* **7,** 1286.

Parry, G., Palmer, D. N., and Williams, D. J. (1976) *FEBS Lett.* **67,** 123.

Paulus, J. (1969) *Anal. Biochem.* **32,** 91.

Perrin, J. H., Vallner, J. J., and Nelson, D. A. (1975) *Biochem. Pharmacol.* **24,** 769.

Pesce, A. J., Rosén, C.-G., and Pasky, T. L. (1971) *Fluorescence Spectroscopy.* Marcel Dekker, New York.

Porschke, D., and Rauh, H. (1983) *Biochemistry* **22,** 4737.

Privalov, P. L., and Khechinashvili, N. N. (1974) *Biopolymers* **7,** 435.

Ptashne, M. (1986) *Nature* **322,** 697.

Quintara, J. R., Lipanov, A. A., and Dickerson, R. E. (1991) *Biochemistry* **30,** 10294.

Raftery, M. A., Dahlquist, F. W., Chan, S. I., and Parsons, S. M. (1968) *J. Biol. Chem.* **243,** 4175.

Rand, R. P., Fuller, N. L., Butko, P., Francis, G., and Nicholls, P. (1993) *Biochemistry* **32,** 5925.

Record, M. T., Jr., Mazur, S. J., Melancon, P., Roe, J.-H., Shaner, S. L., and Unger, L. (1981) *Annu. Rev. Biochem.* **50,** 997.

Richter, P. H., and Eigen, M. (1974) *Biophys. Chem.* **2,** 255.

Riggs, A. D., Bourgeois, S., and Cohn, M. (1970a) *J. Mol. Biol.* **53,** 401.

Riggs, A. D., Suzuki, H., and Beecham, J. M. (1970b) *J. Mol. Biol.* **48,** 67.

Robert, C. H., Gill, S. J., and Wyman, J. (1988) *Biochemistry* **27,** 6829.

Robert, C. H., Colosimo, A., and Gill, S. J. (1989) *Biopolymers* **28,** 1705.

Rodbard, D., Rayford, P. L., Cooper, J. A., and Ross, G. T. (1968) *J. Clin. Endocrinol. Metab.* **28,** 1412.

Rodbard, D., Bridson, W., and Rayford, P. L. (1969) *J. Lab. Clin. Med.* **74,** 770.

Roy, K. B., and Miles, H. T. (1982) *Biochemistry* **21,** 57.

Royer, C. A., Smith, W. R., and Beecham, J. M. (1990) *Anal. Biochem.* **191,** 287.

Saroff, H. A. (1991) *Biochemistry* **30,** 10085.

Scatchard, G. (1949) *Ann. N.Y. Acad. Sci.* **51,** 660.

Schnarr, M., and Daune, M. (1984) *FEBS Lett.* **171,** 207.

Schranner, R., and Richter, P. H. (1978) *Biophys. Chem.* **8,** 135.

Sculley, M. J., Nichol, L. W., and Winzor, D. J. (1981) *J. Theor. Biol.* **90,** 365.

Senear, D. F., and Brenowitz, M. (1991) *J. Biol. Chem.* **266,** 13661.

Shah, V. P., Wallace, S. M., and Reigelman, S. (1974) *J. Pharm. Sci.* **63,** 1364.

Shanley, B. C., Clarke, K., and Winzor, D. J. (1985) *Biochem. Pharmacol.* **34,** 141.

Shearwin, K. E., and Winzor, D. J. (1988a) *Arch. Biochem. Biophys.* **260,** 532.

Shearwin, K. E., and Winzor, D. J. (1988b) *Biophys. Chem.* **31,** 287.

Shearwin, K. E., and Winzor, D. J. (1990) *Arch. Biochem. Biophys.* **282,** 297.

Shimer, G. H., Jr., Woody, A.-Y.M., and Woody, R. W. (1988) *Biochim. Biophys. Acta* **950,** 354.

Singer, S. J. (1965) in *The Proteins* (Neurath, H., ed.), vol. 3, p. 65. Academic Press, New York.

Sklar, L. A., Hudson, B. S., and Simoni, R. D. (1977) *Biochemistry* **16,** 5100.

Smith, G. D. (1977) *J. Theor. Biol.* **69,** 275.

Smith, R. F., and Briggs, D. R. (1950) *J. Phys. Colloid Chem.* **54,** 33.

Sophianopoulos, J. A., Durham, S. J., Sophianopoulos, A. J., Ragsdale, H. L., and Cropper, W. P. (1978) *Arch. Biochem. Biophys.* **187,** 132.

Spector, A. A., John, K., and Fletcher, J. E. (1969) *J. Lipid Res.* **10,** 56.

Spotswood, T. M., Evans, J. M., and Richards, J. H. (1967) *J. Am. Chem. Soc.* **89,** 5052.

Steinberg, I. Z., and Schachman, H. K. (1966) *Biochemistry* **5,** 3728.

Steinhardt, J., and Reynolds, J. A. (1969) *Multiple Equilibria in Proteins*, p. 182. Academic Press, New York.

Svensson, H. (1946) *Ark. Kemi Mineral. Geol.* **22A,** No. 10.

Swaisgood, H. E., and Chaiken, I. M. (1987) in *Analytical Affinity Chromatography* (Chaiken, I. M., ed.), p. 65. CRC Press, Boca Raton, FL.

Sykes, B. D. (1969) *J. Am. Chem. Soc.* **91,** 949.

Takeo, K., and Nakamura, S. (1972) *Arch. Biochem. Biophys.* **153,** 1.

Tanford, C. (1950) *J. Am. Chem. Soc.* **72,** 441.

Tanford, C. (1961) *Physical Chemistry of Macromolecules*, p. 548. Wiley, New York.

Tanford, C., Hauenstein, J. D., and Rands, D. G. (1955) *J. Am. Chem. Soc.* **77**, 6409.

Tellam, R., and Winzor, D. J. (1980) *Arch. Biochem. Biophys.* **201**, 20.

Tellam, R., Winzor, D. J., and Nichol, L. W. (1978) *Biochem. J.* **173**, 185.

Tellam, R., de Jersey, J., and Winzor, D. J. (1979) *Biochemistry* **18**, 5316.

Thomas, E. W. (1966) *Biochem. Biophys. Res. Commun.* **24**, 611.

Thompson, C. J., and Klotz, I. M. (1971) *Arch. Biochem. Biophys.* **147**, 178.

Ts'o, P. O. P., Melvin, I. S., and Olson, A. C. (1963) *J. Am. Chem. Soc.* **85**, 1289.

Uehara, Y., Tonomura, B., and Hiromi, K. (1978) *J. Biochem. (Tokyo)* **84**, 1195.

Ueng, T.-H., and Bronner, F. (1979) *Arch. Biochem. Biophys.* **197**, 205.

van Amerongen, H., Kwa, S. L. S., and van Grondelle, R. (1990) *J. Mol. Biol.* **216**, 717.

Van Holde, K. E., and Baldwin, R. L. (1958) *J. Phys. Chem.* **62**, 734.

Van Holde, K. E., and Rossetti, G. P. (1967) *Biochemistry* **6**, 2189.

Velick, S. F., Hayes, J. E., Jr., and Harting, J. (1953) *J. Biol. Chem.* **203**, 527.

Waltham, M. C., Holland, J. W., Nixon, P. F., and Winzor, D. J. (1988) *Biochem. Pharmacol.* **37**, 541.

Ward, L. D., and Winzor, D. J. (1981) *Arch. Biochem. Biophys.* **209**, 650.

Ward, L. D., and Winzor, D. J. (1983) *Biochem. J.* **215**, 685.

Ward, L. D., Howlett, G. J., Hammacher, A., Weinstock, J., Yasukawa, K., Simpson, R. J., and Winzor, D. J. (1995) *Biochemistry* **34**, 2901.

Weber, G. (1975) *Adv. Protein Chem.* **29**, 1.

Weber, G. (1992) *Protein Interactions*. Chapman and Hall, New York.

Williams, B. A., Chervenak, M. C., and Toone, E. J. (1992) *J. Biol. Chem.* **267**, 22907.

Wills, P. R., and Winzor, D. J. (1992) in *Analytical Ultracentrifugation in Biochemistry and Polymer Science* (Harding, S. E., Rowe, A. J., and Horton, J. C., eds.), p. 311. Roy. Soc. Chem., Cambridge, England.

Wills, P. R., Comper, W. D., and Winzor, D. J. (1993) *Arch. Biochem. Biophys.* **300**, 206.

Wilson, C. J., Bogoyevitch, M. A., and Winzor, D. J. (1990) *Biochem. Pharmacol.* **40**, 1672.

Winzor, D. J. (1985) in *Affinity Chromatography: A Practical Approach* (Dean, P. D. G., Johnson, W. S., and Middle, F. A., eds.), p. 149. IRL Press, Oxford.

Winzor, D. J. (1991) *Biochim. Biophys. Acta* **1115**, 141.

Winzor, D. J. (1992) *J. Chromatogr.* **597**, 67.

Winzor, D. J. (1995) in *Affinity Separations: A Practical Approach* (Matejtschuk, P., ed.) (in press), Oxford University Press, Oxford.

Winzor, D. J., and de Jersey, J. (1989) *J. Chromatogr.* **492**, 377.

Winzor, D. J., and Jackson, C. M. (1993) in *Handbook of Affinity Chromatography* (Kline, T., ed.), p. 253. Marcel Dekker, New York.

Winzor, D. J., and Scheraga, H. A. (1963) *Biochemistry* **2**, 1263.

Winzor, D. J., and Scheraga, H. A. (1964) *J. Phys. Chem.* **68**, 338.

Winzor, D. J., and Wills, P. R. (1995) *Biophys. Chem.* (in press).

Winzor, D. J., Ward, L. D., and Nichol, L. W. (1982) *J. Theor. Biol.* **98,** 171.

Winzor, D. J., Stevens, A., and Augusteyn, R. C. (1989) *Arch. Biochem. Biophys.* **268,** 221.

Winzor, D. J., Bowles, M. R., Pentel, P. R., Schoof, D. D., and Pond, S. M. (1991a) *Mol. Immunol.* **28,** 995.

Winzor, D. J., Nagy, J. A., and Scheraga, H. A. (1991b) *J. Protein Chem.* **10,** 629.

Winzor, D. J., Munro, P. D., and Jackson, C. M. (1992) *J. Chromatogr.* **597,** 57.

Wiseman, T., Williston, S., Brandts, J. F., and Lin, L.-N. (1989) *Anal. Biochem.* **179,** 131.

Wyman, J. (1964) *Adv. Protein Chem.* **19,** 223.

Zeder-Lutz, G., Altschuh, D., Geysen, H. M., Trifilieff, E., Sommermeyer, G., and Van Regenmortel, M. H. V. (1993) *Mol. Immunol.* **30,** 145.

INDEX

Printed in the United States
19865LVS00002B/157